ESSAYS

IN

❄ EUGENICS. ❄

BY

SIR FRANCIS GALTON, F.R.S.

SCOTT-TOWNSEND PUBLISHERS
Washington D.C.

First Published 1909 by the Eugenics Education
Society, London, England

ISBN 1-878465-20-1

Reprinted 1996 by Scott-Townsend Publishers,
P.O. Box 34070, N.W. Washington D.C. 20043
Tel: (202) 371-2700 Fax: (202) 371-1523

INTRODUCTION: SIR FRANCIS GALTON

Sir Francis Galton (1822-1911) ranks among the greatest names in science alongside scholars such as Archimedes, Pythagoras, Leonardo da Vinci and the most prominent of modern academicians. As a geographer, he conducted explorations in Africa; as a meteorologist, he identified anti-cyclones and his pioneering work in forecasting provided the basis for all subsequent weather maps; as a mathe-matician, he developed the use of coefficients of correlation and advanced statistical techniques to the level of calculus; as a biologist and sociologist he showed how statistics could be applied to the study of man; and as an anthropologist he used composite photographs to reveal the median appearance of a population, and developed fingerprinting as a means of identifying individuals. His research into biometrics proved invaluable to paleontology, and his techniques for measuring human abilities contributed heavily to subsequent psychological research, besides demonstrating the importance of heredity in determining human abilities. It was his discoveries in this latter area, and his application of the concept of evolution to the study of modern human populations, that ultimately inspired the concept of sociobiology and – even more importantly – led Galton himself to found the science of Eugenics.

Born near Birmingham on the February 16th, 1822, Francis Galton was but one of a highly talented family, which included among its members his cousin Charles Darwin, the prime originator of the concept of evolution. Educated at King's College, London, Galton graduated from Trinity College, Cambridge, in 1844, and also studied at Birmingham General Hospital and St. George's Hospital, London.

In early life he travelled widely in the Balkans, the Levant, Egypt and the Sudan; but his initial reputation was founded on his explorations in southwest Africa in 1850-52, for which he was awarded the Royal Geographical Society's Gold Medal. In 1856, he was elected a Fellow of the Royal Society.

Entering the British civil service he devoted his early life to the study of geography and meteorology, publishing *Meteorographica* in 1863. But stimulated by his cousin's publication of *The Origin of Species by Means of Natural Selection, or the Preservation of Favoured Races in the Struggle for Life* (1859[1]), Galton turned his mind to

---

[1] London: Murray.

investigating the relevance of evolution as a causal factor explaining the diversity which characterized living men and women, both as breeding populations and as individuals. Mendelian genetics was still unknown at that time, but utilizing statistical methods (his constant advice to researchers was "count") and the laws of probability he laid the foundations for modern behavioral genetics. This he achieved by showing how it was possible to quantify human mental traits and then to apply statistical methods to the data to determine the extent to which heredity and environment affected behavior.

Galton developed procedures for measuring memory span, reaction time and the ability to judge such qualities as length, weight, height, color, touch and smell. As a member of a successful family, Galton was not a poor man, and he conceived of the novel idea of establishing an Anthropometric Laboratory at the International Health Exhibition which opened in 1884 at the South Kensington Science Museum in London, which he financed and supervised. This laboratory collected family histories and biometric data on more than 17,000 individuals, as well as intergenerational data where this was available. Galton himself traced the pedigrees of numerous eminent families in an effort to measure the significance of heredity.

Although it was not then possible positively to distinguish identical from dizygotic twins, Galton studied data relating to the developmental history of twins who were alike at birth and compared these with twins that were unlike at birth. Based on 35 pairs of twins which were judged to be alike at birth, he found that their similarities persisted throughout their subsequent lives, despite changes in environment. By contrast, 20 pairs of unlike twins showed no evidence of growing more alike, even when their environments were similar. Galton may thus be regarded as the founder of twin studies, which in recent decades have become so important for determining the importance of heredity in human behavior and character. He concluded that nature was more important than nurture, and devoted himself thereafter to establishing and advancing the science of eugenics.

For classificatory purposes Galton distinguished fourteen levels of human intellectual ability, ranging from idiocy to genius. Based on a Gaussian curve of the British Population of 1860, Galton determined that a supremely elite group comprising very highly talented "illustrious" individuals, numbered roughly one in a million, while his second category of "eminent" people occurred roughly one in every four thousand of the British population. The destiny of humanity lay in the adequate procreation of this small but highly talented portion of the population which was responsible for virtually all major human advances.

Two of Galton's articles on this topic, published in 1865 in *Macmillan's Magazine*, on the subject of hereditary talent and character formed the basis of his epoch-making *Hereditary Genius: An Enquiry into its Laws and Qualities* (1869[2]). This provided evidence that close relatives of the most distinguished men were much more likely to have achieved distinction themselves than more distant relatives. Thus, 26% of the fathers of the 100 most distinguished persons he had identified were themselves likely to be classed as eminent, as were 23% of their brothers, and 36% of their sons. The 1000 men he identi-fied as being "eminent" proved to be members of just 300 families.

Galton asked himself whether environment, not heredity, could be responsible for this startling pattern, but concluded that heredity was the more important factor on the basis of prevailing evidence. Thus, he observed that many of the men identified had achieved

---

[2] London: Macmillan.

distinction despite the fact that they had risen from relatively poor and unprivileged backgrounds. Furthermore, the adopted kinfolk of Roman Catholic Popes, who benefited from the highest socioeconomic advantages, fared poorly in comparison to the biological sons of eminent men. Galton was strongly critical of Christianity on the grounds that it ignored – or even opposed – the implications of evolution, preferring to teach the doctrine of instant creation and original sin. The truth was that evolution had progressively advanced mankind from the most primitive of origins; but evolution involved constant change – sometimes for the better and sometimes for the worse. Unfortunately modern conditions threatened to reverse the course of human evolution, and nations needed to adopt wise policies with respect to procreation to protect their stock from genetic deterioration. Christianity was promoting genetic decay in many ways, such as the Roman Catholic Church's requirement that its priests be celibate and the tendency of the Anglican church of his day to discourage University dons from marrying.

As is well known, it was Galton who coined the term "eugenics", meaning literally of good birth or well-born. As he explained it, "the processes of evolution are in constant and spontaneous activity, some towards the bad, some towards the good. Our part is to watch for opportunities to intervene by checking the former and giving free rein to the latter." The opposite term, "dysgenics" came into use to refer to forces which reversed the hitherto generally upward course of evolution by promoting the replication of deleterious genes. This term was widely popularized in the present century by another major benefactor of mankind,[3] the Nobel Prize-winning developer of the transistor, William Shockley.

Subsequent to *Hereditary Genius*, other of Francis Galton's more notable publications included *English Men of Science* (1874[4]), *Enquiries into the Human Faculty and Its Development* (1883[5]), *Record of Family Faculties* (1883), *Natural Inheritance* (1889[6]), *Finger Prints* (1892[7]), *Noteworthy Families* (1906[8]), *Memories of My Life* (1908[9]), and *Essays on Eugenics* (1909[10]).

---

[3] See *Shockley on Eugenics and Race* by Roger Pearson (Introduction by Arthur Jensen), Scott-Townsend Publishers, washington D.C. 1992.

[4] London: Macmillan.

[5] London: Macmillan.

[6] London: Macmillan.

[7] London: Macmillan.

[8] London: J. Murray.

[9] London: Methuen.

INTRODUCTION

Knighted in 1909, Sir Francis Galton devoted the last two decades of his life to encouraging informed public opinion to recognize the importance of eugenic policies and to nurturing eugenic research for the benefit of posterity. In 1904 he endowed a Research Fellowship in National Eugenics for the purpose of studying "agencies under social control that may improve or impair the racial qualities of future generations either physically or mentally". He also made major bequests to the University of London to endow a Chair in Eugenics, and to the Eugenics Society, which had come into being with his encouragement and financial assistance.

Francis Galton died at Haslemere in Surrey, England, on January 17, 1911 – happily before he would have been obliged to witness the dysgenic devastation of World War I, which effectively destroyed almost every hope he had entertained for the genetic future of his countrymen. Sadly he himself had no children, and so the English gene pool was permanently weakened by the loss of those of his genes which had made him so outstanding a scholar. His amazing accom-plishments are most fully recorded in the three-volume biography, *The Life, Letters and Labours of Francis Galton*, written

---

[10] London: The Eugenics Education Society.

by his friend, colleague and co-worker Karl Pearson, and published between the years 1914 and 1930.[11]

Galton's *Essays in Eugenics*, which are reprinted here, summarize his conclusions as to the importance of heredity and the need for eugenic measures to counteract the dysgenic influences already beginning to affect the more advanced nations of the world. Handicapped though his perceptions were by the fact that he had no access to the knowledge that genetic science has since made available to us, Galton's conclusions have been largely substantiated by subsequent research. A perusal of his *Essays in Eugenics* not only serves to remind us of our duty to transmit a sound biological heritage to posterity, but also provides a fascinating example of how the prime founder of statistical method applied simple logic to the task of unlocking the secrets of Nature – with remarkable success, notwithstanding the severely restricted quantity of data available in his day.

Roger Pearson

---

[11] London: Cambridge University Press.

PREFACE.

The following Essays are re-printed in the chronological order of their delivery. They will, therefore, help to show something of the progress of Eugenics during the last few years, and to explain my own views upon its aims and methods, which often have been, and still sometimes are, absurdly misrepresented. The practice of Eugenics has already obtained a considerable hold on popular estimation, and is steadily acquiring the status of a practical question, and not that of a mere vision in Utopia.

The power by which Eugenic reform must chiefly be effected, is that of Popular Opinion, which is amply strong enough for that purpose whenever it shall be roused. Public Opinion has done as much as this on many past occasions and in various countries, of which much evidence is given in the Essay on Restrictions in Marriage. It is now ordering our acts more intimately than we are apt to suspect, because the dictates of Public Opinion become so thoroughly assimilated that they seem to be original and individual to those who are guided by them.   By comparing the current ideas at widely different epochs and under widely different civilizations we are able to ascertain what part of our convictions is really innate and permanent, and what part has been acquired and is transient.

It is above all things needful for the successful progress of Eugenics that its advocates should move discreetly and claim no more efficacy on its behalf than the future will confirm ; otherwise a re-action will be invited.   A great deal of investigation is still needed to shew the limit of practical Eugenics, yet enough has been already determined to justify large efforts to instruct the public in an authoritative way, as to the results hitherto obtained by sound reasoning, applied to the undoubted facts of social experience.

My best thanks are due to the Editor of Nature, to the Council of the Sociological Society, and to the Clarendon Press of Oxford, for permission to reprint those among the following essays that first appeared in their Publications.

FRANCIS GALTON.

CONTENTS.

STANDARD SCHEME OF DESCENT

PARENTAL GRADES NUMBER IN EACH	*u* 22	*t* 67	*s* 161	*T* 250	*R* 250	*S* 161	*T* 67	*U* 22
1000 COUPLES BOTH PARENTS OF SAME GRADE AND ONE ADULT MALE CHILD TO EACH								
REGRESSION OF PARENTAL TO FILIAL CENTRES								
22 CHILDREN OF *u*	6	8	6	2				
67 „ or *t*	7	17	23	15	4	1		
161 „ or *s*	5	22	50	52	25	6	1	
250 „ or *T*	2	14	51	86	68	25	4	
250 „ or R		4	25	68	86	51	14	2
161 „ or S		1	6	25	52	50	22	5
67 „ or T			1	4	15	23	17	7
22 „ or U					2	6	8	6
SUMS	20	66	162	252	252	162	66	20

THE POSSIBLE IMPROVEMENT OF THE HUMAN BREED,

Under the Existing Conditions of Law and Sentiment.*

In fulfilling the honourable charge that has been entrusted to me of delivering the Huxley lecture, I shall endeavour to carry out what I understand to have been the wish of its founders, namely, to treat broadly some new topic belonging to a class in which Huxley himself would have felt a keen interest, rather than to expatiate on his character and the work of his noble life.

That which I have selected for to-night is one which has occupied my thoughts for many years, and to which a large part of my published inquiries have borne a direct though silent reference. Indeed, the remarks I am about to make would serve as an additional chapter to my books on " Hereditary Genius " and on " Natural Inheritance." My subject will be the possible improvement of the human race under the existing conditions of law and sentiment. It has not hitherto been ap-

---

* The second Huxley Lecture of the Anthropological Institute, delivered by Francis Galton, D.C.L., D.Sc., F.R.S., on October 29, 1901.

proached along the ways that recent knowledge has laid open, and it occupies in consequence a less dignified position in scientific estimation than it might. It is smiled at as most desirable in itself and possibly worthy of academic discussion, but absolutely out of the question as a practical problem. My aim in this lecture is to show cause for a different opinion. Indeed I hope to induce anthropologists to regard human improvement as a subject that should be kept openly and squarely in view, not only on account of its transcendent importance, but also because it affords excellent but neglected fields for investigation. I shall show that our knowledge is already sufficient to justify the pursuit of this perhaps the grandest of all objects, but that we know less of the conditions upon which success depends than we might and ought to ascertain. The limits of our knowledge and of our ignorance will become clearer as we proceed.

Human Variety.—The natural character and faculties of human beings differ at least as widely as those of the domesticated animals, such as dogs and horses, with whom we are familiar. In disposition some are gentle and good-tempered, others surly and vicious ; some are courageous, others timid ; some are eager, others sluggish ; some have large powers of endurance, others are quickly fatigued ; some are muscular and powerful, others are weak ; some are intelligent, others stupid ; some have tenacious memories of places and persons, others frequently stray and are slow at recognising. The number and variety of aptitudes, especially in dogs, is truly remarkable; among the most notable being the tendency to herd sheep, to point and to retrieve. So it is with the various natural qualities that go towards the making of civic worth in man. Whether it be in character, disposition, energy, intellect, or physical power, we each receive at our birth a definite endowment, allegorised by the parable related in St. Matthew, some receiving many talents, others few; but each person being responsible for the profitable use of that which has been entrusted to him.

Distribution of Qualities in a Nation.— Experience shows that while talents are distributed in endless different degrees, the frequency of those different degrees follows certain statistical laws, of which the best known is the Normal Law of Frequency. This is the result whenever variations are due to the combined action of many small and different causes, whatever may be the causes and whatever the object in which the variations occur, just as twice 2 always makes 4, whatever the objects may be. It therefore holds true with approximate precision for variables of totally different sorts, as, for instance, stature of man, errors made by astronomers in judging minute intervals of time, bullet marks around the bull's-eye in target practice, and differences of marks gained by candidates at competitive examinations. There is no mystery about the fundamental principles of this abstract law; it rests on such simple fundamental conceptions as, that if we toss two pence in the air they will, in the long run, come down one head and one tail twice as often as both heads or both tails. I will assume then, that the talents, so to speak, that go to the formation of civic worth are distributed with rough approximation according to this familiar law. In doing so, I in no way disregard the admirable work of Prof. Karl Pearson on the distribution of qualities, for which he was adjudged the Darwin Medal of the Royal Society a few years ago. He has amply proved that we must not blindly trust the Normal Law of Frequency; in fact, that when variations are minutely studied they rarely fall into that perfect symmetry about the mean value which is one of its consequences. Nevertheless, my conscience is clear in using this law in the way I am about to. I say that *if* certain qualities vary normally, such and such will be the results; that these qualities are of a class that are found, whenever they have been tested, to vary normally to a fair degree of approximation, and consequently we may infer that our results are trustworthy indications of real facts.

A talent is a sum whose exact value few of us care to know, although we all appreciate the inner sense of the beautiful parable. I will, therefore, venture to adapt the phraseology of the allegory to my present purpose by substituting for " talent " the words " normal-talent." The value of this normal talent in respect to each and any specified quality or faculty is such that one-quarter of the people receive for their respective shares more than one normal-talent *over and above* the average of all the shares. Our normal-talent is therefore identical with what is technically known as the " probable error." Therefrom the whole of the following table starts into life, evolved from that of the " *probability integral.*"

TABLE I.—*Normal distribution (to the nearest per ten-thousand and to the nearest per hundred.)*

	$-4°$	$-3°$	$-2°$	$-1°$	M	$+1°$	$+2°$	$+3°$	$+4°$		
v and below	u	t	s	r		R	S	T	U	V and above.	Total
35	180	672	1613	2500		2500	1613	672	180	35	10,000
	2		7	16	25		25	16	7	2	100

It expresses the distribution of any normal quality, or any group of normal qualities, among 10,000 persons in terms of the normal-talent. The M in the upper line occupies the position of Mediocrity, or that of the average of what all have received : the $+1°$, $+2°$, etc., and the $-1°$, $-2°$, etc., refer to normal talents. These numerals stand as graduations at the heads of the vertical lines by which the table is divided. The entries between the divisions are the numbers per 10,000 of those who receive sums between the amounts specified by those divisions. Thus, by the hypothesis, 2500 receive more than M but less than $M + 1°$, 1613 receive more than $M + 1°$ but less than $M + 2°$, and so on. The terminals have only an inner limit, thus 35 receive more than $4°$, some to perhaps a very large and indefinite amount. The divisions might have been carried much farther, but the numbers in the classes between them would become less and less trustworthy. The left half of the series exactly reflects the right half. As it will be useful henceforth to distinguish these classes, I have used the *capital* or large letters R, S, T, U, V, for those above mediocrity and corresponding *italic* or smal letters, *r*, *s*, *t*, *u*, *v*, for those below mediocrity, *r* being the counterpart of R, *s* of S, and so on.

In the lowest line the same values are given, but more roughly, to the nearest whole percentage.

It will assist in comprehending the values of different grades of civic worth to compare them with the corresponding grades of adult male stature in our nation. I will take the figures from my " Natural Inheritance," premising that the distribution of stature in various peoples has been well investigated and shown to be closely normal. The average height of the adult males, to whom my figures refer, was nearly 5 feet 8 inches, and the value of their " normal-talent " (which is a measure of the spread of distribution) was very nearly $1\frac{3}{4}$ inches. From these data it is easily reckoned that Class U would contain men whose heights exceed 6 feet $1\frac{1}{4}$ inches. Even they are tall enough to overlook a hatless mob, while the higher classes, such as V, W and X, tower above it in an increasingly marked degree. So the civic worth (however that term may be defined) of U-class men, and still more of V-class, are notably superior to the crowd, though they are far below the heroic order. The rarity of a V-class man in each specified quality or group of qualities is as 35 in 10,000, or say, for the convenience of using round numbers, as 1 to 300. A man of the W class is ten times rarer, and of the X class rarer still ; but I shall avoid giving any more exact definition of X than as a value considerably rarer than V. This gives a general but just idea of the distribution throughout a population of each and every quality taken separately so far as it is normally distributed. As already mentioned, it does the same for *any* group of normal qualities ; thus, if marks for classics and for mathematics were severally normal in their distribution, the combined marks gained by each candidate in both those subjects would be distributed normally also, this being one of the many interesting properties of the law of frequency.

Comparison of the Normal Classes with those of Mr. Booth.—Let us now compare the normal classes with those into which Mr. Charles Booth has divided the population of all London in a way that corresponds not unfairly with the ordinary conception of grades of civic worth. He reckons them from the lowest upwards, and gives the numbers in each class for East London. Afterwards he treats all London in a similar manner, except that sometimes he combines two classes into one and gives the joint result. For my present purpose, I had to couple them somewhat differently, first disentangling them as I best could. There seemed no better way of doing this than by assigning to the members of each couplet the same proportions that they had in East London. Though this was certainly not accurate, it is probably not far wrong. Mr. Booth has taken unheard of pains in this great work of his to arrive at accurate results, but he emphatically says that his classes cannot be separated sharply from one another. On the contrary, their frontiers blend, and this justifies me in taking slight liberties with his figures. His class A consists of criminals, semi-criminals, loafers and some others, who are in number at the rate of 1 per cent. in all London—that is 100 per 10,000, or nearly three times as many as the v class: they therefore include the whole of v and spread upwards into the u. His class B consists of very poor persons who subsist on casual

TABLE II.—*Comparison of Mr. Booth's Classification of All London with the Normal Classes.*

Nos.	Mr. Booth's classes.	Approx.	Resorted.	Approx.	Nos.	Normal classes.
97	H. All above G	100	100	100	89	T and above
200	{ G. Lower Middle; F. High-class labour above 30s. per week }	200	{ 150; 50 }	150	161	S
382	E. Regular standard earnings from 22s. to 30s. per week	400	{ 200; 200 }	250; 250	250; 250	R; r
227	{ D. Regular earnings under 22s. per week; C. Intermittent earnings, improvident, poor }	200	{ 50; 150 }	150	161	s
94	{ B. Casual; very poor; A. Criminals, loafers, &c.... }	100	100	100	89	t and below
1000			1000	1000	1000	

The two columns headed "Nos." give respectively the numbers per thousand in Mr. Booth's and in the Normal Classes.

earnings, many of whom are inevitably poor from shiftlessness, idleness or drink. The numbers in this and the A class combined closely correspond with those in *t* and all below *t*.

Class C are supported by intermittent earnings; they are a hard-working people, but have a very bad character for improvidence and shiftlessness. In Class D the earnings are regular, but at the low rate of twenty-one shillings or less a week, so none of them rise above poverty, though none are very poor. D and C together correspond to the whole of *s* combined with the lower fifth of *r*. The next class, E, is the largest of any, and comprises all those with regular standard earnings of twenty-two to thirty shillings a week. This class is the recognised field for all forms of co-operation and combination; in short for trades unions. It corresponds to the upper four-fifths of *r*, combined with the lower four-fifths of R. It is therefore essentially the mediocre class, standing as far below the highest in civic worth as it stands above the lowest class with its criminals and semi-criminals. Next above this large mass of mediocrity comes the honourable class *F*, which consists of better paid artisans and foremen. These are able to provide adequately for old age, and their sons become clerks and so forth. *G* is the lower middle class of shop-keepers, small employers, clerks and subordinate professional men, who as a rule are hard-working, energetic and sober.　F and G combined correspond to the upper fifth of R and the whole of S, and are, therefore, a counterpart to D and C.　All above G are put together by Mr. Booth into one class H, which corresponds to our T, U, V and above, and is the counterpart of his two lowermost classes, A and B.　So far, then, as these figures go, civic worth is distributed in fair approximation to the normal law of frequency.　We also see that the classes *t, u, v* and below are undesirables.

Worth of Children.—The brains of the nation lie in the higher of our classes.　If such people as would be classed W or X could be distinguishable as children and procurable by money in order to be reared as Englishmen, it would be a cheap bargain for the nation to buy them at the rate of many hundred or some thousands of pounds per head.　Dr. Farr, the eminent statistician, endeavoured to estimate the money worth of an average baby born to the wife of an Essex labourer and thenceforward living during the usual time and in the ordinary way of his class.　Dr. Farr, with accomplished actuarial skill, capitalised the value at the child's birth of two classes of events, the one the cost of maintenance while a child and when helpless through old age, the other its earnings as boy and man.　On balancing the two sides of the account the value of the baby was found to be five pounds.　On a similar principle, the worth of an X-class baby would be reckoned in thousands of pounds. Some such "talented" folk fail, but most succeed, and many succeed greatly. They found great industries, establish vast undertakings, increase the wealth of multitudes and amass large fortunes for themselves. Others, whether they be rich or poor, are the guides and light of the nation, raising its tone, enlightening its difficulties and imposing its ideals. The great gain that England received through the immigration of the Huguenots would be insignificant to what she would derive from an annual addition of a few hundred children of the classes W and X. I have tried, but not yet succeeded to my satisfaction, to make an approximate estimate of the worth of a child at birth according to the class he is destined to occupy when adult. It is an eminently important subject for future investigators, for the amount of care and cost that might profitably be expended in improving the race clearly depends on its result.

Descent of Qualities in a Population.—Let us now endeavour to obtain a correct understanding of the way in which the varying qualities of each generation are derived from those of its predecessor. How many, for example, of the V class in the offspring come respectively from the V, U, T, S and other classes of parentage ? The means of calculating this question for a normal population are given fully in my " Natural Inheritance."

There are three main senses in which the word parentage might be used. They differ widely, so the calculations must be modified accordingly. (1) The amount of the quality or faculty in question may be known in each parent. (2) It may be known in only one parent. (3) The two parents may belong to the same class, a V-class father in the scale of male classification always marrying a V-class mother, occupying identically the same position in the scale of female classification.

I select this last case to work out as being the one with which we shall here be chiefly concerned. It has the further merit of escaping some tedious preliminary details about converting female faculties into their corresponding male equivalents, before men and women can be treated statistically on equal terms. I shall assume in what follows that we are dealing with an ideal population, in which all marriages are equally fertile, and which is statistically the same in successive generations both in numbers and in qualities, so many per cent. being always this, so many always that, and so on. Further, I shall take no notice of offspring who die before they reach the age of marriage, nor shall I regard the slight numerical inequality of the sexes, but will simply suppose that each parentage produces one couplet of grown-up filials, an adult man and an adult woman.

The result is shown to the nearest whole per thousand in the table up to " V and above,"

TABLE III.—*Descent of Qualities in a Population. (The difference between the sexes only affects the value of the Unit of the Scale of Distribution.)*

Conditions.—(1) Parents to be always alike in class, (2) Statistics of population to continue unchanged, (3) Normal Law of Frequency to be applicable throughout.

	v	u	t	s	r	R	S	T	U	V	Totals
Per 100 Father (or *Mothers*).	2		7	16	25	25	16	7	2		100
Per 10,000 ,, ,,	35	180	671	1614	2500	2500	1614	672	180	35	10,000
Names of classes	v	u	t	s	r	R	S	T	U	V	Totals
Sons / Daughters } of 35 {Fathers / Mothers} of class V ...						1	6	12	10	6	Sons (or *daughters*) 35
,, 180 ,, ,, U ...					4	20	52	61	33	10	180
,, 671 ,, ,, T ...				7	44	150	234	170	57	10	672
,, 1614 ,, ,, S ...			6	57	253	512	509	224	47	5	1613
,, 2500 ,, ,, R ...		3	42	248	678	860	510	140	18	3	2502
,, 2500 ,, ,, r ...	3	18	140	510	860	678	248	42	3		2502
,, 1614 ,, ,, s ...	5	47	224	509	512	253	57	6			1613
,, 671 ,, ,, t ...	10	57	170	234	150	44	7				672
,, 180 ,, ,, u ...	10	33	61	52	20	4					180
,, 35 ,, ,, v ...	6	10	12	6	1						35
Total 10,000 Fathers (or *Mothers*)	34	168	655	1623	2522	2522	1623	655	168	34	10,004
,, 100 ,, ,, ...	2		7	16	25	25	16	7	2		

Note.—The agreement in distribution between fathers (or *mothers*) and sons (or *daughters*) is exact to the nearest whole per centage. The slight discrepancy in the ten-thousandths is mainly due to the classes being too few and too wide ; theoretically they should be extremely numerous and narrow.

to the nearest ten thousands. They may be read either as applying to fathers and their sons when adult, or to mothers and their daughters when adult, or, again, to parentages and filial couplets. I will not now attempt to explain the details of the calculation to those to whom these methods are new. Those who are familiar with them will easily understand the exact process from what follows. There are three points of reference in a scheme of descent which may be respectively named " mid-parental," " genetic " and " filial " centres. In the present case of both parents being alike, the position of the mid-parental centre is identical with that of either parent separately. The position of the filial centre is that from which the children disperse. The genetic centre occupies the same position in the parental series that the filial centre does in the filial series. " Natural Inheritance " contains abundant proof, both observational and theoretical, that the genetic centre is not and cannot be identical with the parental centre, but is always more mediocre, owing to the combination of ancestral influences—which are generally mediocre— with the purely parental ones. It also shows that the regression from the parental to the genetic centre, in the case of stature at least, would amount to two-thirds under the conditions we are now supposing. The regression is indicated in the diagram used to illustrate this paper, by converging lines which are directed towards the same point below, but are stopped at one-third of the distance on the way to it. The contents of each parental class are supposed to be concentrated at the foot of the median axis of that class, this being the vertical line that divides its contents into equal parts. Its position is approximately, but not exactly, half-way between the divisions that bound it, and is as easily calculated for the extreme classes, which have no outer terminals, as for any of the others. These median points are respectively taken to be the positions of the parental centres of the whole of each of the classes ; therefore the positions attained by the converging lines that proceed from them at the points where they are stopped, represent the genetic centres. From these the filials disperse to the right and left with a " spread " that can be shown to be three-quarters that of the parentages. Calculation easily deter-mines the number of the filials that fall into the class in which the filial centre is situated, and of those that spread into the classes on each side. When the parental contributions from all the classes to each filial class are added together they will express the distribu-tion of the quality among the whole of the offspring. Now it will be observed in the table that the numbers in the classes of the offspring are identical with those of the parents, when they are reckoned to the nearest whole percentage, as should be the case according to the hypothesis. Had the classes been narrower and more numerous, and if the calculations had been carried on to two more places of decimals, the correspondence would have been identical to the nearest ten-thousandth. It was unnecessary to take the trouble of doing this, as the table affords a sufficient basis for what I am about to say. Though it does not profess to be more than approximately true in detail, it is certainly trustworthy in its general form, including as it does the effects of regression, filial dispersion, and the equation that connects a parental generation with a filial one when they are statistically alike. Minor corrections will be hereafter required, and can be applied when we have a better knowledge of the material. In the meantime it will serve as a standard table of descent from each generation of a people to its successor.

Economy of Effort.—I shall now use the table to show the economy of concentrating our attention upon the highest classes. We will therefore trace the origin of the V class—which is the highest in the table. Of its 34 or 35 sons, 6 come from V parentages, 10 from U, 10 from T, 5 from S, 3 from R, and none from any class below R. But the numbers of the contributing parentages have also to be taken into account. When this is done, we see that the lower classes make their scores owing to their quantity and not to their quality; for while 35 V-class parents suffice to produce 6 sons of the V class, it takes 2500 R-class fathers to produce 3 of them. Consequently the richness in produce of V-class parentages is to that of the R-class in an inverse ratio, or as 143 to 1. Similarly, the richness in produce of V-class children from parentages of the classes U, T, S, respectively, is as 3, $11\frac{1}{2}$ and 55, to 1. Moreover, nearly one-half of the produce of V-class parentages are V or U taken together, and nearly three-quarters of them are either V, U or T. If then we desire to increase the output of V-class offspring, by far the most profitable parents to work upon would be those of the V-class, and in a threefold less degree those of the U class.

When both parents are of the V class the quality of parentages is greatly superior to those in which only one parent is a V. In that case the regression of the genetic centre goes twice as far back towards mediocrity, and the spread of the distribution among filials becomes nine-tenths of that among the parents, instead of being only three-quarters. The effect is shown in table IV.

There is a difference of fully two divisions in the position of the genetic centre, that of the single V parentage being only a trifle nearer mediocrity than that of the double T. Hence it would be bad economy to spend much effort in furthering marriages with a higher class on only one side.

TABLE IV.—*Distribution of sons. (1) One parent of class V., the other unknown. (2) Both parents of class V (from Table II., with decimal point and an 0).*

	Distribution of Sons								Total
	t	s	r	R	S	T	U	V	
One V-parent......	0·3	1·2	3·5	7·9	9·6	7·5	3·6	1·3	34·3
Two V-parents ...				3·0	5·0	10·0	10·0	6·0	34·0

Position of the filial centre of (1) = 1·44, of (2) = 2·89. When both parents are T it = 1·58.

Marriage of like to like.—In each class of society there is a strong tendency to inter-marriage, which produces a marked effect in the richness of brain power of the more cultured families. It produces a still more marked effect of another kind at the lowest step of the social scale, as will be painfully evident from the following extracts from the work of Mr. C. Booth (i. 38), which refer to his Class A, who form, as has been said, the lowermost third of our " v and below." "Their life is the life of savages, with vicissitudes of extreme hardship and occasional excess. From them come the battered figures who slouch through the streets and play the beggar or the bully. They render no useful service, they create no wealth ; more often they destroy it. They degrade whatever they touch, and as individuals are perhaps incapable of improvement . . . but I do not mean to say that there are not individuals of every sort to be found in the mass. Those who are able to wash the mud may find some gems in it. There are at any rate many very piteous cases. Whatever doubt there may be as to the exact numbers of this class, it is certain that they bear a very small proportion to the rest of the population, or even to class B, with which they are mixed up and from which it is at times difficult to separate them. . . . They are barbarians, but they are a handful. . . . ." He says further, " It is much to be desired and to be hoped that this class may become less hereditary in its character; there appears to be no doubt that it is now hereditary to a very considerable extent."

Many who are familiar with the habits of these people do not hesitate to say that it would be an economy and a great benefit to the country if all habitual criminals were resolutely segregated under merciful surveillance and peremptorily denied opportunities for producing offspring. It would abolish a source of suffering and misery to a future generation, and would cause no unwarrantable hardship in this.

Diplomas.—It will be remembered that Mr. Booth's classification did not help us beyond classes higher than S in civic worth. If a strong and widely felt desire should arise to discover young men whose position was of the V, W or X order, there would not be much difficulty in doing so. Let us imagine, for a moment, what might be done in any great University, where the students are in continual competition in studies, in athletics, or in public meetings, and where their characters are publicly known to associates and to tutors. Before attempting to make a selection, acceptable definitions of civic worth would have to be made in alternative terms, for there are many forms of civic worth. The number of men of the V, W or X classes whom the University was qualified to contribute annually must also be ascertained. As was said, the proportion in the general population of the V class to the remainder is as 1 to 300, and that of the W class as 1 in 3000. But students are a somewhat selected body because the cleverest youths, in a scholastic sense, usually find their way to Universities. A considerably high level, both intellectually and physically, would be required as a qualification for candidature. The limited number who had not been automatically weeded away by this condition might be submitted in some appropriate way to the independent votes of fellow-students on the one hand, and of tutors on the other, whose ideals of character and merit necessarily differ. This ordeal would reduce the possible winners to a very small number, out of which an independent committee might be trusted to make the ultimate selection. They would be guided by personal interviews. They would take into consideration all favourable points in the family histories of the candidates, giving appropriate hereditary weight to each. Probably they would agree to pass over unfavourable points, unless they were notorious and flagrant, owing to the great difficulty of ascertaining the real truth about them. Ample experience in making selections has been acquired even by scientific societies, most of which work well, including perhaps the award of their medals, which the fortunate recipients at least are tempted to consider judicious. The opportunities for selecting women in this way are unfortunately fewer, owing to the smaller number of female students between whom comparisons might be made on equal terms. In the selection of women, when nothing is known of their athletic proficiency, it would be especially necessary to pass a high and careful medical examination ; and as their personal qualities do not usually admit of being tested so thoroughly as those of men, it would be necessary to lay all the more stress on hereditary family qualities, including those of fertility and prepotency.

Correlation between Promise in Youth and subsequent Performance.—No serious difficulty seems to stand in the way of classifying and giving satisfactory diplomas to youths of either sex, supposing there were a strong demand for it. But some real difficulty does lie in the question—Would such a classification be a trustworthy forecast of qualities in later life ? The scheme of descent of qualities may hold good between the parents and the offspring at similar ages, but that is not the information we really want. It is the descent of qualities from men to men, not from youths to youths. The accidents that make or mar a career do not enter into the scope of this difficulty. It resides entirely in the fact that the development does not cease at the time of youth, especially in the higher natures, but that faculties and capabilities which were then latent subsequently unfold and become prominent. Putting aside the effects of serious illness, I do not suppose there is any risk of retrogression in capacity before old age comes on. The mental powers that a youth possesses continue with him as a man ; but other faculties and new dispositions may arise and alter the balance of his character He may cease to be efficient in the way of which he gave promise, and he may perhaps become efficient in unexpected directions.

The correlation between youthful promise and performance in mature life has never been properly investigated. Its measurement presents no greater difficulty, so far as I can foresee, than in other problems which have been successfully attacked. It is one of those alluded to in the beginning of this lecture as bearing on race-improvement, and being on its own merits suitable for anthropological inquiry. Let me add that I think its neglect by the vast army of highly educated persons who are connected with the present huge system of competitive examinations to be gross and unpardonable. Neither schoolmasters, tutors, officials of the Universities, nor of the State department of education, have ever to my knowledge taken any serious step to solve this important problem, though the value of the present elaborate system of examinations cannot be rightly estimated until it is solved. When the value of the correlation between youthful promise and adult performance shall have been determined, the figures given in the table of descent will have to be reconsidered.

Augmentation of Favoured Stock.—The possibility of improving the race of a nation depends on the power of increasing the productivity of the best stock. This is far more important than that of repressing the productivity of the worst. They both raise the average, the latter by reducing the undesirables, the former by increasing those who will become the lights of the nation. It is therefore all important to prove that favour to selected individuals might so increase their productivity as to warrant the expenditure in money and care that would be necessitated. An enthusiasm to improve the race would probably express itself by granting diplomas to a select class of young men and women, by encouraging their intermarriages, by hastening the time of marriage of women of that high class, and by provision for rearing children healthily. The means that might be employed to compass these ends are dowries, especially for those to whom moderate sums are important, assured help in emergencies during the early years of married life, healthy homes, the pressure of public opinion, honours, and above all the introduction of motives of religious or quasi-religious character. Indeed, an enthusiasm to improve the race is so noble in its aim that it might well give rise to the sense of a religious obligation. In other lands there are abundant instances in which religious motives make early marriages a matter of custom, and continued celibacy to be regarded as a disgrace, if not a crime. The customs of the Hindoos, also of the Jews, especially in ancient times, bear this out. In all costly civilisations there is a tendency to shrink from marriage on prudential grounds. It would, however, be possible so to alter the conditions of life that the most prudent course for an X class person should lie exactly opposite to its present direction, for he or she might find that there were advantages and not disadvantages in early marriage, and that the most prudent course was to follow the natural instincts.

We have now to consider the probable gain in the number and worth of adult offspring to these favoured couples. First as regards the effect of reducing the age at marriage. There is unquestionably a tendency among cultured women to delay or even to abstain from marriage; they dislike the sacrifice of freedom and leisure, of opportunities for study and of cultured companionship. This has to be reckoned with. I heard of the reply of a lady official of a College for Women to a visitor who inquired as to the after life of the students. She answered that one-third profited by it, another third gained little good, and a third were failures. " But what become of the failures ? " " Oh, they marry."

There appears to be a considerable difference between the earliest age at which it is physiologically desirable that a woman should marry and that at which the ablest, or at least the most cultured, women usually do. Acceleration in the time of marriage, often amounting to 7 years, as from 28 or 29 to 21 or 22, under influences such as those mentioned above, is by no means improbable. What would be its effect on productivity ? It might be expected to act in two ways :—

(1) By shortening each generation by an amount roughly proportionate to the diminution in age at which marriage occurs. Suppose the span of each generation to be shortened by one-sixth, so that six take the place of five, and that the productivity of each marriage is unaltered, it follows that one-sixth more children will be brought into the

world during the same time, which is, roughly equivalent to increasing the productivity of an unshortened generation by that amount.

(2) By saving from certain barrenness the earlier part of the child-bearing period of the woman. Authorities differ so much as to the direct gain of fertility due to early marriage that it is dangerous to express an opinion. The large and thriving families that I have known were the offspring of mothers who married very young.

The next influence to be considered is that of healthy homes. These and a simple life certainly conduce to fertility. They also act indirectly by preserving lives that would otherwise fail to reach adult age. It is not necessarily the weakest who perish in this way, for instance, zymotic disease falls indiscriminately on the weak and the strong.

Again, the children would be healthier and therefore more likely in their turn to become parents of a healthy stock. The great danger to high civilisations, and remarkably so to our own, is the exhaustive drain upon the rural districts to supply large towns. Those who come up to the towns may produce large families, but there is much reason to believe that these dwindle away in subsequent generations. In short, the towns sterilise rural vigour.

As one of the reasons for choosing the selected class would be that of hereditary fertility, it follows that the selected class would respond more than other classes to the above influences.

I do not attempt to appraise the strength of the combined six influences just described. If each added one-sixth to the produce the number of offspring would be doubled. This does not seem impossible considering the large families of colonists, and of those in many rural districts; but it is a high estimate. Perhaps the fairest approximation may be that these influences would cause the X women to bring into the world an average of one adult son and one adult daughter *in addition* to what they would otherwise have produced. The table of descent applies to one son or to one daughter per couple; it may now be read as specifying the net gain and showing its distribution. Should this estimate be thought too high, the results may be diminished accordingly.

It is no absurd idea that outside influences should hasten the age of marrying and make it customary for the best to marry the best. A superficial objection is sure to be urged that the fancies of young people are so incalculable and so irresistible that they cannot be guided. No doubt they are so in some exceptional cases. I lately heard from a lady who belonged to a county family of position that a great aunt of hers had scandalised her own domestic circle two generations ago by falling in love with the undertaker at her father's funeral and insisting on marrying him. Strange vagaries occur, but considerations of social position and of fortune, with frequent opportunities of intercourse, tell much more in the long run than sudden fancies that want roots. In a community deeply impressed with the desire of encouraging marriages between persons of equally high ability, the social pressure directed to produce the desired end would be so great as to ensure a notable amount of success.

Profit and Loss.—The problem to be solved now assumes a clear shape. A child of the X class (whatever X signifies) would have been worth so and so at its birth, and one of each of the other grades respectively would have been worth so and so ; 100 X parentages can be made to produce a net gain of 100 adult sons and 100 adult daughters who will be distributed among the classes according to the standard table of descent. The total value of the prospective produce of the 100 parentages can then be estimated by an actuary, and consequently the sum that it is legitimate to spend in favouring an X parentage. The clear and distinct statement of a problem is often more than half way towards its solution. There seems no reason why this one should not be solved between limiting values that are not too wide apart to be useful.

Existing Activities.—Leaving aside profitable expenditure from a purely money point of view, the existence should be borne in mind of immense voluntary activities that have nobler aims. The annual voluntary contributions in the British Isles to public charities alone amount, on the lowest computation, to fourteen million pounds, a sum which Sir H. Burdett asserts on good grounds is by no means the maximum obtainable.*

("Hospitals and Charities," 1898, p. 85.)

There are other activities long since existing which might well be extended. I will not dwell, as I am tempted to do, on the endowments of scholarships and the like, which aim at finding and educating the fittest youths for the work of the nation ; but I will refer to that wholesome practice during all ages of wealthy persons interesting themselves in and befriending poor but promising lads. The number of men who have owed their start in a successful life to help of this kind must have struck every reader of biographies. This relationship of befriender and befriended

---

*The 80 charitable bequests of and exceeding £9000, made in 1808 alone, amounted to more than 3½ millions of pounds. (Whitaker's Almanack to 1909, p. 433).

"It being far more humane to *prevent* suffering than to *alleviate* it after it has occurred, why will not charitably disposed persons leave substantial sums of money to the furtherance of Eugenic Study and practice, and of popularising the result ? The money would be well bestowed." *Francis Galton, 1909.*

I learn on high legal authority that the form of bequest which would be most appropriate in present circumstances, and be free from the pit-falls that lie in the way of charitable bequests, is " I bequeath to my trusted friend A.B., of ....., absolutely, the sum of £...... in the hope and confidence that he will apply the same in furtherance of Eugenic Study and practice, but without imposing on him any trust or legal obligation so to do." F.G.

is hardly to be expressed in English by a simple word that does not connote more than is intended. The word "patron" is odious. Recollecting Dr. Johnson's abhorrence of the patrons of his day, I turned to an early edition of his dictionary in hope of deriving some amusement as well as instruction from his definition of the word, and I was not disappointed. He defines "patron" as "a wretch who supports with insolence and is repaid with flattery." That is totally opposed to what I would advocate, namely, a kindly and honourable relation between a wealthy man who has made his position in the world and a youth who is avowedly his equal in natural gifts, but who has yet to make it. It is one in which each party may well take pride and I feel sure that if its value were more widely understood it would become commoner than it is.

Many degrees may be imagined that lie between mere befriendment and actual adoption, and which would be more or less effective in freeing capable youths from the hindrances of narrow circumstances; in enabling girls to marry early and suitably, and in securing favour for their subsequent offspring. Something in this direction is commonly but half unconsciously done by many great landowners whose employments for man and wife, together with good cottages, are given to exceptionally deserving couples. The advantage of being connected with a great and liberally managed estate being widely appreciated, there are usually more applicants than vacancies, so selection can be exercised. The consequence is that the class of men found upon these properties is markedly superior to those in similar positions elsewhere. It might well become a point of honour, and as much an avowed object, for noble families to gather fine specimens of humanity around them, as it is to procure and maintain fine breeds of cattle and so forth, which are costly, but repay in satisfaction.

There is yet another existing form of princely benevolence which might be so extended as to exercise a large effect on race improvement. I mean the provision to exceptionally promising young couples of healthy and convenient houses at low rentals. A continually renewed settlement of this kind can be easily imagined, free from the taint of patronage, and analogous to colleges with their self-elected fellowships and rooms for residence, that should become an exceedingly desirable residence for a specified time. It would be so in the same way that a good club by its own social advantages attracts desirable candidates. The tone of the place would be higher than elsewhere, on account of the high quality of the inmates, and it would be distinguished by an air of energy, intelligence, health and self-respect and by mutual helpfulness.

Prospects.—It is pleasant to contrive Utopias, and I have indulged in many, of which a great society is one, publishing intelligence and memoirs, holding yearly elections, administering large funds, establishing personal relations like a missionary society with its missionaries, keeping elaborate registers and discussing them statistically with honest precision. But the first and pressing point is to thoroughly justify any crusade at all in favour of race improvement. More is wanted in the way of un-biased scientific inquiry along the many roads I have hurried over, to make every stepping-stone safe and secure, and to make it certain that the game is really worth the candle. All I dare hope to effect by this lecture is to prove that in seeking for the improvement of the race we aim at what is apparently possible to accomplish, and that we are justified in following every path in a resolute and hopeful spirit that seems to lead towards that end. The magnitude of the inquiry is enormous, but its object is one of the highest man can accomplish. The faculties of future generations will necessarily be distributed according to laws of heredity, whose statistical effects are no longer vague, for they are measured and expressed in formulæ. We cannot doubt the existence of a great power ready to hand and capable of being directed with vast benefit as soon as we shall have learnt to understand and to apply it.   To no nation is a high human breed more necessary than to our own, for we plant our stock all over the world and lay the foundation of the dispositions and capacities of future millions of the human race.

EUGENICS: ITS DEFINITION, SCOPE AND AIMS.*

Eugenics is the science which deals with all influences that improve the inborn qualities of a race ; also with those that develop them to the utmost advantage. The improvement of the inborn qualities, or stock, of some one human population, will alone be discussed here.

What is meant by improvement ? What by the syllable *Eu* in Eugenics, whose English equivalent is *good* ? There is considerable difference between goodness in the several qualities and in that of the character as a whole. The character depends largely on the *proportion* between qualities whose balance may be much influenced by education. We must therefore leave morals as far as possible out of the discussion, not entangling ourselves with the almost hopeless difficulties they raise as to whether a character as a whole is good or bad. Moreover, the goodness or badness of character is not absolute, but relative to the current form of civilisation. A fable will best explain what is meant. Let the scene be the Zoological Gardens in the quiet hours of

*Read before the Sociological Society at a Meeting in the School of Economics and Political Science (London University), on May 16th, 1904. Professor KARL PEARSON, F.R.S., in the chair.

the night, and suppose that, as in old fables, the animals are able to converse, and that some very wise creature who had easy access to all the cages, say a philosophic sparrow or rat, was engaged in collecting the opinions of all sorts of animals with a view of elaborating a system of absolute morality. It is needless to enlarge on the contrariety of ideals between the beasts that prey and those they prey upon, between those of the animals that have to work hard for their food and the sedentary parasites that cling to their bodies and suck their blood, and so forth. A large number of suffrages in favour of maternal affection would be obtained, but most species of fish would repudiate it, while among the voices of birds would be heard the musical protest of the cuckoo. Though no agreement could be reached as to absolute morality, the essentials of Eugenics may be easily defined. All creatures would agree that it was better to be healthy than sick, vigorous than weak, well fitted than ill-fitted for their part in life. In short that it was better to be good rather than bad specimens of their kind, whatever that kind might be. So with men. There are a vast number of conflicting ideals of alternative characters, of incompatible civilisations ; but all are wanted to give fulness and interest to life. Society would be very dull if every man resembled the highly estimable Marcus Aurelius or Adam Bede. The aim of Eugenics is to represent each class or sect by its best specimens ; that done, to leave them to work out their common civilisation in their own way.

A considerable list of qualities can be easily compiled that nearly every one except "cranks" would take into account when picking out the best specimens of his class. It would include health, energy, ability, manliness and courteous disposition. Recollect that the natural differences between dogs are highly marked in all these respects, and that men are quite as variable by nature as other animals in their respective species. Special aptitudes would be assessed highly by those who possessed them, as the artistic faculties by artists, fearlessness of inquiry and veracity by scientists, religious absorption by mystics, and so on. There would be self-sacrificers, self-tormentors and other exceptional idealists, but the representatives of these would be better members of a community than the body of their electors. They would have more of those qualities that are needed in a State, more vigour, more ability, and more consistency of purpose. The community might be trusted to refuse representatives of criminals, and of others whom it rates as undesirable.

Let us for a moment suppose that the practice of Eugenics should hereafter raise the average quality of our nation to that of its better moiety at the present day and consider the gain. The general tone of domestic, social and political life would be higher. The race as a whole would be less foolish, less frivolous, less excitable and politically more provident than now. Its demagogues who "played to the gallery" would play to a more sensible gallery than at present. We should be better fitted to fulfil our vast imperial opportunities. Lastly, men of an order of ability which is now very rare, would become more frequent, because the level out of which they rose would itself have risen.

The aim of Eugenics is to bring as many influences as can be reasonably employed, to cause the useful classes in the community to contribute *more* than their proportion to the next generation.

The course of procedure that lies within the functions of a learned and active Society such as the Sociological may become, would be somewhat as follows :—

1. Dissemination of a knowledge of the laws of heredity so far as they are surely known, and promotion of their farther study. Few seem to be aware how greatly the knowledge of what may be termed the *actuarial* side of heredity has advanced in recent years. The *average* closeness of kinship in each degree now admits of exact definition and of being treated mathematically, like birth and death-rates, and the other topics with which actuaries are concerned.

2. Historical inquiry into the rates with which the various classes of society (classified according to civic usefulness) have contributed to the population at various times, in

ancient and modern nations.   There is strong reason for believing that national rise and decline is closely connected with this influence. It seems to be the tendency of high civilisation to check fertility in the upper classes, through numerous causes, some of which are well known, others are inferred, and others again are wholly obscure.   The latter class are apparently analogous to those which bar the fertility of most species of wild animals in zoological gardens.   Out of the hundreds and thousands of species that have been tamed, very few indeed are fertile when their liberty is restricted and their struggles for livelihood are abolished ; those which are so and are otherwise useful to man becoming domesticated.   There is perhaps some connection between this obscure action and the disappearance of most savage races when brought into contact with high civilization, though there are other and well-known concomitant causes. But while most barbarous races disappear, some, like the negro, do not.   It may therefore be expected that types of our race will be found to exist which can be highly civilised without losing fertility ; nay, they may become more fertile under artificial conditions, as is the case with many domestic animals.

3.   Sytematic collection of facts showing the circumstances under which large and thriving families have most frequently originated ; in other words, the *conditions* of Eugenics.   The names of the thriving families

in England have yet to be learnt, and the conditions under which they have arisen. We cannot hope to make much advance in the science of Eugenics without a careful study of facts that are now accessible with difficulty, if at all. The definition of a thriving family, such as will pass muster for the moment at least is one in which the children have gained distinctly superior positions to those who were their class-mates in early life. Families may be considered "large" that contain not less than three adult male children. It would be no great burden to a Society including many members who had Eugenics at heart, to initiate and to preserve a large collection of such records for the use of statistical students. The committee charged with the task would have to consider very carefully the form of their circular and the persons entrusted to distribute it. The circular should be simple, and as brief as possible, consistent with asking all questions that are likely to be answered truly, and which would be important to the inquiry. They should ask, at least in the first instance, only for as much information as could be easily, and would be readily, supplied by any member of the family appealed to. The point to be ascertained is the *status* of the two parents at the time of their marriage, whence its more or less eugenic character might have been predicted, if the larger knowledge that we now hope to obtain had then existed. Some account would, of course, be wanted of their race, profession, and residence ; also of their own respective parentages, and of their brothers and sisters.   Finally, the reasons would be required why the children deserved to be entitled a " thriving " family, to distinguish worthy from unworthy success. This manuscript collection might hereafter develop into a " golden book " of thriving families.   The Chinese, whose customs have often much sound sense, make their honours retrospective.   We might learn from them to show that respect to the parents of noteworthy children, which the contributors of such valuable assets to the national wealth richly deserve.   The act of systematically collecting records of thriving families would have the further advantage of familiarising the public with the fact that Eugenics had at length become a subject of serious scientific study by an energetic Society.

4.   Influences affecting Marriage.   The remarks of Lord Bacon in his essay on Death may appropriately be quoted here.   He says with the view of minimising its terrors :

" There is no passion in the mind of men so weak but it mates and masters the fear of death. - Revenge triumphs over death ; love slights it ; honour aspireth to it ; grief flyeth to it ; fear pre-occupateth it."

Exactly the same kind of considerations apply to marriage. The passion of love seems so overpowering that it may be thought folly to try to direct its course. But plain facts do not confirm this view. Social influences of all kinds have immense power in the end, and they are very various. If unsuitable marriages from the Eugenic point of view were banned socially, or even regarded with the unreasonable disfavour which some attach to cousin-marriages, very few would be made. The multitude of marriage restrictions that have proved prohibitive among uncivilised people would require a volume to describe.

5.   Persistence in setting forth the national importance of Eugenics.   There are three stages to be passed through. *Firstly* it must be made familiar as an academic question, until its exact importance has been understood and accepted as a fact ; *Secondly* it must be recognised as a subject whose practical development deserves serious consideration ; and *Thirdly* it must be introduced into the national conscience, like a new religion. It has, indeed, strong claims to become an orthodox religious tenet of the future, for Eugenics co-operates with the workings of Nature by securing that humanity shall be represented by the fittest races. What Nature does blindly, slowly, and ruthlessly, man may do providently, quickly, and kindly. As it lies within his power, so it becomes his duty to work in that direction ; just as it is his duty to succour neighbours who suffer misfortune. The improvement of our stock seems to me one of the highest objects that we can reasonably attempt. We are ignorant of the ultimate destinies of

humanity, but feel perfectly sure that it is as noble a work to raise its level in the sense already explained, as it would be disgraceful to abase it.   I see no impossibility in Eugenics becoming a religious dogma among mankind, but its details must first be worked out sedulously in the study.   Over-zeal leading to hasty action would do harm, by holding out expectations of a near golden age, which will certainly be falsified and cause the science to be discredited.   The first and main point is to secure the general intellectual acceptance of Eugenics as a hopeful and most important study.   Then let its principles work into the heart of the nation, who will gradually give practical effect to them in ways that we may not wholly foresee.

RESTRICTIONS IN MARRIAGE.*

It is proposed in the following remarks to meet an objection that has been repeatedly urged against the possible adoption of any system of Eugenics, namely, that human nature would never brook interference with the freedom of marriage.

In my reply, I shall proceed on the not unreasonable assumption, that when the subject of Eugenics shall be well understood, and when its lofty objects shall have become generally appreciated, they will meet with some recognition both from the religious sense of the people and from its laws. The question now to be considered is, how far have marriage restrictions proved effective, when sanctified by the religion of the time, by custom, and by law? I appeal from arm-chair criticism to historical facts.

To this end, a brief history will be given of a few widely spread customs. It will be seen that with scant exceptions they are based on social expediency, and not on natural instincts. Each of the following paragraphs might have been expanded into a long chapter

Read before the Sociological Society, on Tuesday, February 14th, at a meeting in the School of Economics and Political Science (University of London), Clare Market, W.C., Dr. E. WESTERMARCK in the Chair.

had that seemed necessary. Those who desire to investigate the subject further can easily do so by referring to standard works in anthropology, among the most useful of which, for the present purpose, are Frazer's *Golden Bough*, Westermarck's *History of Marriage*, Huth's *Marriage of Near Kin*, and Crawley's *Mystic Rose*.

1. MONOGAMY. It is impossible to label mankind by one general term, either as animals who instinctively take a plurality of mates, or who consort with only one, for history suggests the one condition as often as the other. Probably different races, like different individuals, vary considerably in their natural instincts. Polygamy may be understood either as having a plurality of wives; or, as having one principal wife and many secondary but still legitimate wives, or any other recognised but less legitimate connections; in one or other of these forms it is now permitted —by religion, customs, and law—to at least one-half of the population of the world, though its practice may be restricted to a few, on account of cost, domestic peace, and the insufficiency of females. Polygamy holds its ground firmly throughout the Moslem world. It exists throughout India and China in modified forms, and it is entirely in accord with the sentiments both of men and women in the larger part of negro Africa. It was regarded as a matter of course in the early Biblical days. Jacob's twelve children were born of four

mothers all living at the same time, namely, Leah, and her sister Rachel, and their respective handmaids Billah and Zilpah. Long afterwards, the Jewish kings emulated the luxurious habits of neighbouring potentates and carried polygamy to an extreme degree. For Solomon, see I Kings xi. 3. For his son Rehoboam, see II Chron. xi. 21. The history of the subsequent practice of the custom among the Jews is obscure, but the Talmud contains no law against polygamy, It must have ceased in Judæa by the time of the Christian Era. It was not then allowed in either Greece or Rome. Polygamy was unchecked by law in profligate Egypt, but a reactionary and ascetic spirit existed, and some celibate communities were formed in the service of Isis, who seem to have exercised a large though indirect influence in introducing celibacy into the early Christian Church. The restriction of marriage to one living wife subsequently became the religion and the law of all Christian nations, though licence has been widely tolerated in royal and other distinguished families, as in those of some of our English kings. Polygamy was openly introduced into Mormonism by Brigham Young, who left seventeen wives, and fifty-six children. He died in 1877 ; polygamy was suppressed soon after *(Encyc. Brit.*, xvi. 827.)

It is unnecessary for my present purpose to go further into the voluminous data connected with marriages such as these in all parts of the world. Enough has been said to show that the prohibition of polygamy, under severe penalties by civil and ecclesiastical law, has been due not to any natural instinct against the practice, but to consideration of social well-being. I conclude that equally strict limitations to freedom of marriage might, under the pressure of worthy motives, be hereafter enacted for Eugenic and other purposes.

2. ENDOGAMY, or the custom of marrying exclusively *within* one's own tribe or caste, has been sanctioned by religion and enforced by law, in all parts of the world, but chiefly in long settled nations where there is wealth to bequeath and where neighbouring communities profess different creeds. The details of this custom, and the severity of its enforcement, have everywhere varied from century to century. It was penal for a Greek to marry a barbarian, for a Roman patrician to marry a plebeian, for a Hindu of one caste to marry one of another caste, and so forth. Similar restrictions have been enforced in multitudes of communities, even under the penalty of death.

A very typical instance of the power of law over the freedom of choice in marriage, and which was by no means confined to Judæa, is that known as the Levirate. It shows that family property and honour were once held by the Jews to dominate over individual preferences. The Mosaic law actually *compelled* a man to marry the widow of his brother if he left no male issue. (Deuteron. xxv.) Should the brother refuse, " then shall his brother's wife come unto him in the presence of the elders, and loose his shoe from off his foot, and spit in his face; and she shall answer and say, so shall it be done unto the man that doth not build up his brother's house. And his name shall be called in Israel the house of him that hath his shoe loosed." The form of this custom survives to the present day and is fully described and illustrated under the article " Halizah" (= taking off, untying) in the *Jewish Cyclopædia.* Jewish widows are now almost invariably re-married with this ceremony. They are as we might describe it, " given away " by a kinsman of the deceased husband, who puts on a shoe of an orthodox shape which is kept for the purpose, the widow unties the shoe, spits, but now on the *ground*, and repeats the specified words.

The duties attached to family property led to the history, which is very strange to the ideas of the present day, of Ruth's advances to Boaz under the advice of her mother. " It came to pass at midnight " that Boaz "was startled (see marginal note in the Revised Version) and turned himself, and behold a woman lay at his feet," who had come in " softly and uncovered his feet and laid her down." He told her to lie still until the early morning and then to go away. She returned home and told her mother, who said, "Sit still, my daughter, until thou know how the matter will fall, for the man will not rest until he have finished the thing this day." She was right. Boaz took legal steps to disembarrass himself of the claims of a still nearer kinsman, "who drew off his shoe"; so Boaz married Ruth. Nothing could be purer from the point of view of those days, than the history of Ruth. The feelings of the modern social world would be shocked if the same thing were to take place now in England.

Evidence from the various customs relating to endogamy show how choice in marriage may be dictated by religious custom. That is, by a custom founded on a religious view of family property and family descent. Eugenics deal with what is more valuable than money or lands, namely the heritage of a high character, capable brains, fine physique, and vigour; in short, with all that is most desirable for a family to possess as a birthright. It aims at the evolution and preservation of high races of men, and it as well deserves to be strictly enforced as a religious duty, as the Levirate law ever was.

3. EXOGAMY is, or has been, as widely spread as the opposed rule of endogamy just described. It is the duty enforced by custom, religion, and law, of marrying *outside* one's own clan, and is usually in force amongst small and barbarous communities. Its

former distribution is attested by the survival in nearly all countries, of ceremonies based on " marriage by capture." The remarkable monograph on this subject by the late Mr. McLennan is of peculiar interest. It was one of the earliest, and perhaps the most successful, of all attempts to decipher pre-historic customs by means of those now existing among barbarians, and by the marks they have left on the traditional practices of civilised nations, including ourselves. Before his time those customs were regarded as foolish, and fitted only for antiquarian trifling. In small fighting communities of barbarians, daughters are a burden ; they are usually killed while infants, so few women are found in a tribe who were born in it. It may sometimes happen that the community has been recently formed by warriors who brought no women, and who, like the Romans in the old story, could only supply themselves by capturing those of neighbour-ing tribes. The custom of capture grows ; it becomes glorified because each wife is a living trophy of the captor's heroism, and marriage within the tribe soon comes to be considered an unmanly, and at last a shameful act. The modern instances of this among barbarians are very numerous.

4. AUSTRALIAN MARRIAGES. The follow-ing is a brief clue, and apparently a true one, to the complicated marriage restrictions among Australian bushmen, which are enforced

by the penalty of death, and which seem to be partly endogamous in origin and partly otherwise.   The example is typical of those of many other tribes that differ in detail.

A and B are two tribal classes ; 1 and 2 are two other and *independent* divisions of the tribe (probably by totems). Any person, taken at random, is equally likely to have either letter or either numeral by birthright, and his or her numeral and letter are well known to all the community. Hence the members of the tribe are sub-classed into four sub-divisions, A1, A2, B1, B2. The rule is that a man may marry those women only, whose letter and numeral are both different to his own. Thus A1 can marry only B2, the other three sub-divisions A1, A2, and B1 being absolutely barred to him. As to the children, there is a difference of practice in different parts : in the cases most often described, the child takes its father's letter and its mother's numeral, which determines class by paternal descent. In other cases the arrangement runs in the contrary way, that is by maternal descent.

The cogency of this rule is due to custom, religion and law, and is so strong that nearly all Australians would be horrified at the idea of breaking it. If anyone dared to do so, he would probably be clubbed to death.

Here then is another restriction to the freedom of marriage which might with equal propriety have been applied to the furtherance of some form of Eugenics.

5. TABOO. The survival of young animals largely depends on their inherent timidity, their keen sensitiveness to warnings of danger by their parents and others, and to their tenacious recollection of them. It is so with human children, who are easily terrified by nurses' tales and thereby receive more or less durable impressions.

A vast complex of motives can be brought to bear upon the naturally susceptible minds of children, and of uneducated adults who are mentally little more than big children. The constituents of this complex are not sharply distinguishable, but they form a recognisable whole that has not yet received an appropriate name, in which religion, superstition, custom, tradition, law, and authority all have part. This group of motives will for the present purpose be entitled " immaterial " in contrast to material ones. My contention is that the experience of all ages and all nations shows that the immaterial motives are frequently far stronger than the material ones, the relative power of the two being well illustrated by the tyranny of taboo in many instances, called as it is by different names in different places. The facts relating to taboo form a voluminous literature, the full effect of which cannot be conveyed by brief summaries. It shows how, in most parts of the world, acts that are apparently insignificant, have been invested with ideal importance, and how the doing of this or that has been followed by outlawry or death, and how the mere terror of having unwittingly broken a taboo, may suffice to kill the man who broke it. If non-eugenic unions were prohibited by such taboos, none would take place.

6. PROHIBITED DEGREES. The institution of marriage, as now sanctified by religion and safeguarded by law in the more highly civilised nations, may not be ideally perfect, nor may it be universally accepted in future times, but it is the best that has hitherto been devised for the parties primarily concerned, for their children, for home life, and for society. The degrees of kinship within which marriage is prohibited, is with one exception quite in accordance with modern sentiment, the exception being the disallowal of marriage with the sister of a deceased wife, the propriety of which is greatly disputed and need not be discussed here. The marriage of a brother and sister would excite a feeling of loathing among us that seems implanted by nature, but which further inquiry will show, has mainly arisen from tradition and custom.

We will begin by giving due weight to certain assigned motives. (1) Indifference and even repugnance between boys and girls, irrespectively of relationship, who have been reared in the same barbarian home. (2) Close likeness, as between the members of a thorough-bred stock, causes some sexual indifference : thus highly bred dogs lose much of their sexual desire for one another, and are apt to consort with mongrels. (3) Contrast is an element in sexual attraction which has not yet been discussed quantitatively. Great resemblance creates indifference, and great dissimilarity is repugnant. The maximum of attractiveness must lie somewhere between the two, at a point not yet ascertained. (4) The harm due to continued interbreeding has been considered, as I think, without sufficient warrant, to cause a presumed strong natural and instinctive *repugnance* to the marriage of near kin. The facts are that close and continued interbreeding invariably does harm after a few generations, but that a single cross with near kinsfolk is practically innocuous. Of course a sense of repugnance might become correlated with any harmful practice, but there is no evidence that it is *repugnance* with which interbreeding is correlated, but only *indifference;* this is equally effective in preventing it, but is quite another thing. (5) The strongest reason of all in civilised countries appears to be the earnest desire not to infringe the sanctity and freedom of the social relations of a family group, but this has nothing to do with instinctive sexual repugnance. Yet it is through the latter motive alone, so far as I can judge, that we have acquired our apparently instinctive horror of marrying within near degrees.

Next as to facts. History shows that the horror now felt so strongly did not exist in early times. Abraham married his half-sister Sarah, " she is indeed the sister, the daughter of my father, but not the daughter of my mother, and she became my wife." (Gen. xx. 12). Amram, the father of Moses and Aaron, married his aunt, his father's sister Jochabed. The Egyptians were accustomed to marry sisters. It is unnecessary to go earlier back in Egyptian history than to Ptolemies, who, being a new dynasty, would not have dared to make the marriages they did in a conservative country, unless popular opinion allowed it. Their dynasty includes its founder Ceraunus, who is not numbered ; the numbering begins with his son Soter, and goes on to Ptolemy XIII., the second husband of Cleopatra. Leaving out her first husband, Ptolemy XII., as he was a mere boy, and taking in Ceraunus, there are thirteen Ptolemies to be considered. Between them, they contracted eleven incestuous marriages, eight with whole sisters, one with a half-sister, and two with nieces. Of course the object was to keep the royal line pure, as was done by the ancient Peruvians. It would be tedious to follow out the laws enforced at various times and in the various states of Greece during the classical ages. Marriage was at one time permitted in Athens between half-brothers and half-sisters, and the marriage between uncle and niece was thought commendable in the time of Pericles when it was prompted by family considerations.

In Rome the practice varied much, but there were always severe restrictions. Even in its dissolute period, public opinion was shocked by the marriage of Claudius with his niece.

A great deal more evidence could easily be adduced, but the foregoing suffices to prove that there is no instinctive repugnance felt universally by man, to marriage within the prohibited degrees, but that its present strength is mainly due to what I call immaterial considerations. It is quite conceivable that a non-eugenic marriage should hereafter excite no less loathing than that of a brother and sister would do now.

7. CELIBACY. The dictates of religion n respect to the opposite duties of leading celibate lives, and of continuing families, have been contradictory. In many nations it is and has been considered a disgrace to bear no children, and in other nations celibacy has has been raised to the rank of a virtue of the highest order. The ascetic character of the African portion of the early Christian Church, as already remarked, introduced the merits of celibate life into its teaching. During the fifty or so generations that have elapsed since the establishment of Christianity, the nunneries and monasteries, and the celibate lives of Catholic priests, have had vast social effects, how far for good and how far for evil need not be discussed here. The point which I wish to enforce is the potency, not only of the religious sense in aiding or deterring

marriage, but more especially the influence and authority of ministers of religion in enforcing celibacy. They have notoriously used it when aid has been invoked by members of the family on grounds that are not religious at all, but merely of family expediency. Thus, at some times and in some Christian nations, every girl who did not marry while still young, was practically compelled to enter a nunnery from which escape was afterwards impossible.

It is easy to let the imagination run wild on the supposition of a whole-hearted acceptance of Eugenics as a national religion ; that is of the thorough conviction by a nation that no worthier object exists for man than the improvement of his own race ; and when efforts as great as those by which nunneries and monasteries were endowed and maintained should be directed to fulfil an opposite purpose. I will not enter further into this. Suffice it to say, that the history of conventual life affords abundant evidence on a very large scale, of the power of religious authority in directing and withstanding the tendencies of human nature towards freedom in marriage.

CONCLUSION.—Seven different subjects have now been touched upon. They are monogamy, endogamy, exogamy, Australian marriages, taboo, prohibited degrees and celebacy. It has been shown under each of these heads how powerful are the various combinations of immaterial motives upon marriage selection, how they may all become hallowed by religion, accepted as custom and enforced by law. Persons who are born under their various rules live under them without any objection. They are unconscious of their restrictions, as we are unaware of the tension of the atmosphere. The subservience of civilised races to their several religious superstitions, customs, authority, and the rest, is frequently as abject as that of barbarians. The same classes of motives that direct other races, direct ours, so a knowledge of their customs helps us to realise the wide range of what we may ourselves hereafter adopt, for reasons that will be as satisfactory to us in those future times, as theirs are or were to them, at the time when they prevailed.

Reference has frequently been made to the probability of Eugenics hereafter receiving the sanction of religion. It may be asked, " how can it be shown that Eugenics fall within the purview of our own." It cannot, any more than the duty of making provision for the future needs of oneself and family, which is a cardinal feature of modern civilization, can be deduced from the Sermon on the Mount. Religious precepts, founded on the ethics and practice of olden days, require to be reinterpreted to make them conform to the needs of progressive nations. Ours are already so far behind modern requirements that much of our practice and our profession cannot be reconciled without illegitimate casuistry. It seems to me that few things are more needed by us in England than a revision of our religion, to adapt it to the intelligence and needs of the present time. A form of it is wanted that shall be founded on reasonable bases and enforced by reasonable hopes and fears, and that preaches honest morals in unambiguous language, which good men who take their part in the work of the world, and who know the dangers of sentimentalism, may pursue without reservation.

STUDIES IN NATIONAL EUGENICS*

It was stated in the *Times*, January, 26, 1905, that at a meeting of the Senate of the University of London, Mr. Edgar Schuster, M.A., of New College, Oxford, was appointed to the Francis Galton Research Fellowship in National Eugenics. " Mr. Schuster will in particular carry out investigations into the history of classes and families, and deliver lectures and publish memoirs on the subjects of his investigations."

Now that this appointment has been made, it seems well to publish a suitable list of subjects for eugenic inquiry. It will be a programme that binds no one, not even myself, for I have not yet had the advantage of discussing it with others, and may hereafter wish to largely revise and improve what is now provisionally sketched. The use of this paper lies in its giving a general outline of what, according to my present view, requires careful investigation, of course not all at once, but step by step, at possibly long intervals.

---

*Communicated at a meeting of the Sociological Society held in the School of Economics and Political Science (University of London), Clare Market, W.C., on Tuesday, February 14th, at 4 p.m.

I.   Estimation of the average quality of the offspring of married couples, from their personal and ancestral data.   This includes questions of fertility, and the determination of the "probable error" of the estimate for individuals, according to the data employed.

(a) "Biographical Index to Gifted Families," modern and recent, for publication.   It might be drawn up on the same principle as my "Index to Achievements of Near Kinsfolk of some of the Fellows of the Royal Society" (see "Sociological Papers," Vol. I., p. 85).   The Index refers only to facts creditable to the family, and to such of these as have already appeared in publications, which are quoted as authority for the statements.   Other biographical facts that may be collected concerning these families are to be preserved for statistical use only.

(b) Biographies of capable families, who do not rank as "gifted," are to be collected, and kept in MS., for statistical use, but with option of publication.

(c) Biographies of families, who, as a whole, are distinctly below the average in health, mind, or physique, are to be collected.   These include the families of persons in asylums of all kinds, hospitals, and prisons. To be kept for statistical use only.

(d) Parentage and progeny of representatives of each of the social classes of the community, to determine how far each class is derived from, and contributes to, its own and other classes.   This inquiry must be carefully planned beforehand.

(e) Insurance Office data.   An attempt to be made to carry out the suggestions of Mr. Palin Elderton, "Sociological Papers," Vol. I., p. 62, of obtaining material that the authorities would not object to give, and whose discussion might be advantageous to themselves as well as to Eugenics.   The matter is now under consideration, so more cannot be said.

II.   Effects of action by the State and by Public Institutions.

(*f*) Habitual criminals.   Public opinion is beginning to regard with favour the project of a prolonged segregation of habitual criminals, for the purpose of restricting their opportunities for (1) continuing their depredations, and (2) producing low class offspring. The enquiries spoken of above (see *c*) will measure the importance of the latter object.

(*g*) Feeble minded.   Aid given to Institutions for the feeble minded are open to the suspicions that they may eventually promote their marriage and the production of offspring like themselves.   Inquiries are needed to test the truth of this suspicion.

(*h*) Grants towards higher education.   Money spent in the higher education of those who are intellectually unable to profit by it lessens the sum available for those who can do so.   It might be expected that aid systematically given on a large scale to the more capable would have considerable eugenic effect, but the subject is complex and needs investigation.

(*i*) Indiscriminate charity, including out-door relief. There is good reason to believe that the effects of indiscriminate charity are notably non-eugenic.   This topic affords a wide field for inquiry.

III.   Other influences that further or restrain particular classes of marriage.

The instances are numerous in recent times in which social influences have restrained or furthered freedom of marriage.   A judicious selection of these would be useful, and might be undertaken as time admits.   I have myself just communicated to the Sociological Society a memoir entitled " Restrictions in Marriage," in which remarkable instances are given of the dominant power of religion, law and custom.   This will suggest the sort of work now in view, where less powerful influences have produced statistical effects of appreciable amount.

IV. Heredity.

The facts after being collected are to be discussed, for improving our knowledge of the laws both actuarial and of physiological heredity, the recent methods of advanced statistics being of course used. It is possible that a study of the effect on the offspring of differences in the parental qualities may prove important.

It is to be considered whether a study of Eurasians, that is, of the descendants of Hindoo and English parents, might not be advocated in proper quarters, both on its own merits as a topic of national importance and as a test of the applicability of the Mendelian hypotheses to men. Eurasians have by this time intermarried during three consecutive generations in sufficient numbers to yield trustworthy results.

V. Literature.

A vast amount of material that bears on Eugenics exists in print, much of which is valuable and should be hunted out and catalogued. Many scientific societies, medical, actuarial, and others, publish such material from time to time. The experiences of breeders of stock of all kinds, and those of horticulturists, fall within this category.

VI. Co-operation.

After good work shall have been done and become widely recognised, the influence of eugenic students in stimulating others to contribute to their inquiries may become powerful. It is too soon to speculate on this, but every good opportunity should be seized to further co-operation, as well as the knowledge and application of Eugenics.

VII. Certificates.

In some future time, dependent on circumstances, I look forward to a suitable authority issuing Eugenic certificates to candidates for them. They would imply a more than an average share of the several qualities of at least goodness of constitution, of physique, and of mental capacity. Examinations upon which such certificates might be granted are already carried on, but separately; some by the medical advisers of insurance offices, some by medical men as to physical fitness for the army, navy and Indian services, and others in the ordinary scholastic examinations. Supposing constitution, physique and intellect to be three independent variables (which they are not), the men who rank among the upper third of each group would form only one twenty-seventh part of the population. Even allowing largely for the correlation of those qualities, it follows that a moderate severity of selection in each of a few particulars would lead to a severe all-round selection. It is not necessary to pursue this further.

The above brief memorandum does not profess to deal with more than the pressing problems in Eugenics. As that science becomes better known, and the bases on which it rests are more soundly established, new problems will arise, especially such as relate to its practical application. All this must bide its time; there is no good reason to anticipate it now. Of course useful suggestions in the present embryonic condition of Eugenic study would be timely, and might prove very helpful to students.

MR. GALTON'S REPLY

TO REMARKS MADE DURING THE DISCUSSION
THAT FOLLOWED.

This Society has cause to congratulate itself on the zeal and energy which has brought together so large a body of opinion. We have had verbal contributions from four eminent specialists in anthropology: Dr. Haddon, Dr. Mott, Mr. Crawley, and

Dr. Westermarck, and numerous written communications have been furnished by well known persons. At the time that I am revising and extending these words no less than twenty-six contributions to the discussion are in print. Want of space compels me to confine my reply to those remarks that seem more especially to require it, and to do so very briefly, for Eugenics is a wide study, with an uncounted number of side issues into which those who discuss it are tempted to stray. If, however, sure advance is to be made, these issues must be thoroughly explored, one by one, and partial discussion should as far as possible be avoided. To change the simile, we have to deal with a formidable chain of strongholds, which must be severally attacked in force, reduced, and disposed of, before we can proceed freely.

In the first place, it is a satisfaction to find that no one impugns the conclusion which my memoir was written to justify, that history tells how restrictions in marriage, even of an excessive kind, have been contentedly accepted very widely, under the guidance of what I called "immaterial motives." This is all I had in view when writing it.

Certificates.—One of the comments on which I will remark is that if certificates were now offered to those who passed certain examinations into health, physique, moral and intellectual powers, and hereditary gifts, great mistakes would be made by the examiners. I fully agree that it is too early to devise a satisfactory system of marks for giving what might be styled "honour-certificates," because we do not yet possess sufficient data to go upon. On the other hand there are persons who are exceptionally and unquestionably unfit to contribute offspring to the nation, such as those mentioned in Dr. Mott's bold proposals. The best methods of dealing with these are now ripe for immediate consideration.

Breeding for points.—It is objected by many that there cannot be unanimity on the "points" that it is most desirable to breed for. I fully discussed this objection in my memoir read here last spring, showing that some qualities such as health and vigour were thought by all to be desirable, and the opposite undesirable, and that this sufficed to give a first direction to our aims. It is a safe starting point, though a great deal more has to be inquired into as we proceed on our way. I think that some contributors to this discussion have been needlessly alarmed. No question has been raised by me of breeding men like animals for particular points, to the disregard of all-round efficiency in physical, intellectual (including moral), and hereditary qualifications. Moreover, as statistics have shown, the best qualities are largely correlated. The youths who became judges, bishops, statesmen, and leaders of progress in England could have furnished formidable athletic teams in their times.   There is a tale, I know not how far founded on fact, that Queen Elizabeth had an eye to the calves of the legs of those she selected for bishops. There is something to be said in favour of selecting men by their physical characteristics for other than physical purposes. It would decidedly be safer to do so than to trust to pure chance.

The residue.—It is also objected that if the inferior moiety of a race are left to intermarry, their produce will be increasingly inferior. This is certainly an error. The law of "regression towards mediocrity" insures that their offspring as a whole, will be superior to themselves, and if as I sincerely hope, a freer action will be hereafter allowed to selective agencies than hitherto, the portion of the offspring so selected would be better stiil. The influences that now *withstand* the free action of selective agencies are numerous, they include indiscriminate charity.

Passion of love.—The argument has been repeated that love is too strong a passion to be restrained by such means as would be tolerated at the present time. I regret that I did not express the distinction that ought to have been made between its two stages, that of slight inclination and that of falling thoroughly into love, for it is the first of these rather than the second that I hope the popular feeling of the future will successfully resist. Every match-making mother appreciates the difference. If a girl is taught to look upon a class of men as tabooed, whether owing to rank, creed, connections, or other causes, she does not regard them as possible husbands and turns her thoughts elsewhere. The proverbial " Mrs. Grundy " has enormous influence in checking the marriages she considers indiscreet.

Eugenics as a factor in religion.—Remarks have been made concerning eugenics as a religion ; this will be the subject of the brief memoir that follows these remarks.

It is much to be desired that competent persons would severally take up one or other of the many topics mentioned in my second memoir, or others of a similar kind, and work it thoroughly out as they would any ordinary scientific problem ; in this way solid progress would be made. I must be allowed to re-emphasise my opinion that an immense amount of investigation has to be accomplished before a definite system of Eugenics can be safely framed.

EUGENICS AS A FACTOR IN RELIGION.

Eugenics strengthens the sense of social duty in so many important particulars that the conclusions derived from its study ought to find a welcome home in every tolerant religion. It promotes a far-sighted philanthropy, the acceptance of parentage as a serious responsibility, and a higher conception of patriotism. The creed of eugenics is founded upon the idea of evolution; not on a passive form of it, but on one that can to some extent direct its own course. Purely passive, or what may be styled mechanical evolution, displays the awe inspiring spectacle of a vast eddy of organic turmoil, originating we know not how, and travelling we know not whither. It forms a continuous whole from first to last, reaching backward beyond our earliest knowledge and stretching forward as far as we think we can foresee. But it is moulded by blind and wasteful processes, namely, by an extravagant production of raw material and the ruthless rejection of all that is superfluous, through the blundering steps of trial and error. The condition at each successive moment of this huge system, as it issues from the already quiet past and is about to invade the still undisturbed future, is one of violent internal commotion. Its elements are in constant flux and change, though its general form alters but slowly. In this respect it resembles the curious stream of cloud that sometimes seems attached to a mountain top during the continuance of a strong breeze; its constituents are always changing, though its shape as a whole hardly varies. Evolution is in any case a grand phantasmagoria, but it assumes an infinitely more interesting aspect under the knowledge that the intelligent action of the human will is, in some small measure, capable of guiding its course. Man has the power of doing this largely so far as the evolution of humanity is concerned; he has already affected the quality and distribution of organic life so widely that the changes on the surface of the earth, merely through his disforestings and agriculture, would be recognisable from a distance as great as that of the moon.

As regards the practical side of eugenics, we need not linger to re-open the unending argument whether man possesses any creative power of will at all, or whether his will is not also predetermined by blind forces or by intelligent agencies behind the veil, and whether the belief that man can act independently is more than a mere illusion. This matters little in practice, because men, whether fatalists or not, work with equal vigour whenever they perceive they have the power to act effectively.

Eugenic belief extends the function of philanthropy to future generations, it renders its action more pervading than hitherto, by dealing with families and societies in their entirety, and it enforces the importance of the marriage covenant by directing serious attention to the probable quality of the future offspring. It sternly forbids all forms of sentimental charity that are harmful to the race, while it eagerly seeks opportunity for acts of personal kindness, as some equivalent to the loss of what it forbids. It brings the tie of kinship into prominence and strongly encourages love and interest in family and race. In brief, eugenics is a virile creed, full of hopefulness, and appealing to many of the noblest feelings of our nature.

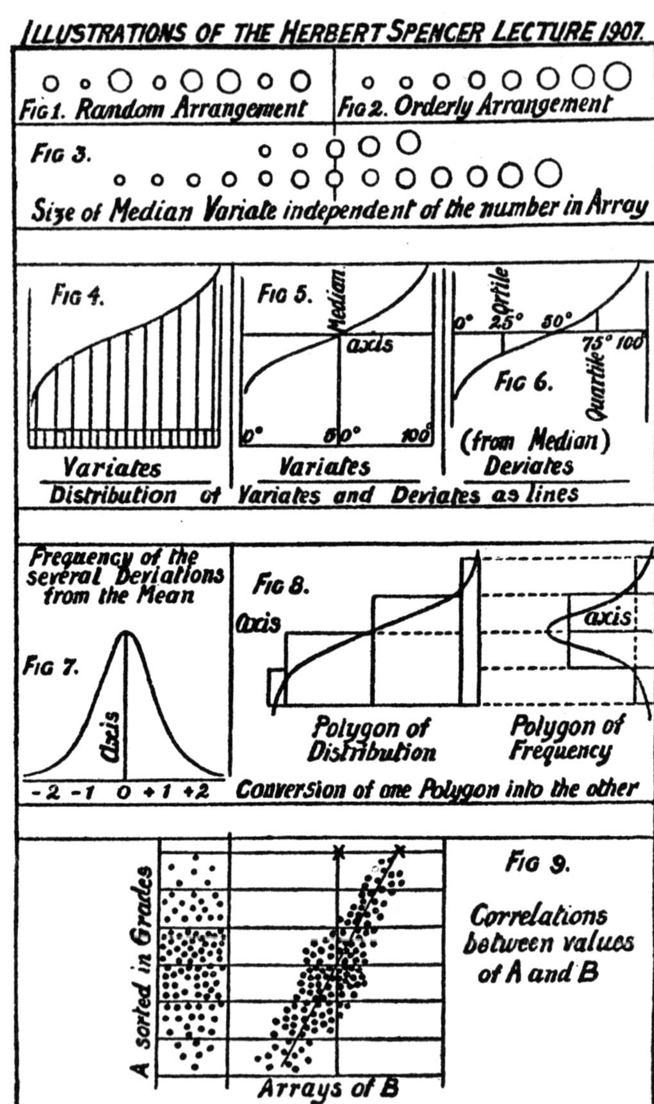

ILLUSTRATIONS OF THE HERBERT SPENCER LECTURE 1907.
Fig 1. Random Arrangement
Fig 2. Orderly Arrangement
Fig 3.
Size of Median Variate independent of the number in Array
Fig 4.
Variates
Fig 5.
Median
axis
0°
50°
100°
Variates
0°
25°
50°
75° 100
Quartile
2ⁿ qrtile
Fig 6.
(from Median)
Deviates
Distribution of Variates and Deviates as lines
Frequency of the several Deviations from the Mean
Fig 7.
Axis
-2 -1 0 +1 +2
Fig 8.
Axis
axis
Polygon of Distribution
Polygon of Frequency
Conversion of one Polygon into the other
A sorted in Grades
Arrays of B
Fig 9.
Correlations between values of A and B

PROBABILITY, THE FOUNDATION OF EUGENICS.*

THE request so honourable to myself, to be the Herbert Spencer lecturer of this year, aroused a multitude of vivid recollections. Spencer's strong personality, his complete devotion to a self-imposed and life-long task, together with rare gleams of tenderness visible amidst a wilderness of abstract thought, have left a unique impression on my mind that years fail to weaken.

I do not propose to speak of his writings ; they have been fully commented on elsewhere, but I desire to acknowledge my personal debt to him, which is large. It lies in what I gained through his readiness to discuss any ideas I happened to be full of at the time, with quick sympathy and keen criticism. It was his custom for many afternoons to spend an hour or two of rest in the old smoking room of the Athenaeum Club, strolling into an adjoining compartment for a game of billiards when the table was free. Day after day on those afternoons I enjoyed brief talks with him, which were often of

---

*The Herbert Spencer Lecture delivered before the University at Oxford, June 5th, 1907.

exceptional interest to myself. All that kind of comfort and pleasure has long ago passed from me. Among the many things of which age deprives us, I regret few more than the loss of contemporaries. When I was young I felt diffident in the presence of my seniors, partly owing to a sense that the ideas of the young cannot be in complete sympathy with those of the old. Now that I myself am old it seems to me that my much younger friends keenly perceive the same difference, and I lose much of that outspoken criticism which is an invaluable help to all who investigate.

History of Eugenics.

It must have surprised you as it did myself to find the new word 'Eugenics' in the title both of the Boyle Lecture, delivered in Oxford about a fortnight ago, and of this. It was an accident, not a deliberate concurrence, and I accept it as a happy omen. The field of Eugenics is so wide that there is no need for myself, the second lecturer, to plant my feet in the footsteps of the first ; on the contrary, it gives freedom by absolving me from saying much that had to be said in one way or another. I fully concur in the views so ably presented by my friend and co-adjutor, Professor Karl Pearson, and am glad to be dispensed from further allusion to subjects that formed a large portion of his lecture, on which he is a far better guide and an infinitely higher authority than myself.

In giving the following sketch of the history of Eugenics I am obliged to be egotistical, because I kindled the feeble flame that struggled doubtfully for a time until it caught hold of adjacent stores of suitable material, and became a brisk fire, burning freely by itself, and again because I have had much to do with its progress quite recently.

The word ' Eugenics ' was coined and used by me in my book *Human Faculty*, published as long ago as 1883, which has long been out of print ; it is, however, soon to be re-published in a cheap form.* In it I emphasized the essential brotherhood of mankind, heredity being to my mind a very real thing; also the belief that we are born to act, and not to wait for help like able-bodied idlers, whining for doles.   Individuals appear to me as finite detachments from an infinite ocean of being, temporarily endowed with executive powers. This is the only answer I can give to myself in reply to the perpetually recurring questions of ' Why ? whence ? and whither ? ' The immediate 'whither?' does not seem wholly dark, as some little information may be gleaned concerning the direction in which Nature, so far as we know of it, is now moving—Namely, towards the evolution of mind, body, and character in increasing energy and co-adaptation.

I have often wondered that the poem of Hyperion, by Keats—that magnificent torso of an incompleted work—has not been placed

---

*Dent's " Everyman's Library," price One Shilling.

in the very forefront of past speculations on evolution. Keats is so thorough that he makes the very Divinities to be its product. The earliest gods such as Coelus, born out of Chaos, are vague entities, they engender Saturn, Oceanus, Hyperion, and the Titan brood, who supersede them. These in their turn are ousted from dominion by their own issue, the Olympian Gods. A notable advance occurs at each successive stage in the quality of the Divinities. When Hyperion, newly terrified by signs of impending overthrow, lies prostrate on the earth ' his ancient mother, for some comfort yet,' the voice of Coelus from the universal space, thus ' whispered low and solemn in his ear . . . yet do thou strive, for thou art capable . . . my life is but the life of winds and tides, no more than winds and tides can I prevail, but thou canst.' I have quoted only disjointed fragments of this wonderful poem, enough to serve as a reminder to those who know it, but will add ten consecutive lines from the speech of the fallen Oceanus to his comrades, which give a summary of evolution as here described :

> As Heaven and Earth are fairer, fairer far
> Than Chaos and black Darkness, though once chiefs,
> And as we show beyond that Heaven and Earth
> In form and shape compact and beautiful,
> In Will, in action free, companionship,
> And thousand other signs of purer life ;
> So on our heels a fresh perfection treads
> A power more strong in beauty, born of us
> And fated to excel us, as we pass
> In glory that old Darkness.

He ends with ' this is the truth, and let it be your balm.' The poem is a noble conception, founded on the crude cosmogony of the ancient Greeks.

The ideas have long held my fancy that we men may be the chief, and perhaps the only executives on earth. That we are detached on active service with, it may be only illusory, powers of free-will. Also that we are in some way accountable for our success or failure to further certain obscure ends, to be guessed as best we can. That though our instructions are obscure they are sufficiently clear to justify our interference with the pitiless course cf Nature, whenever it seems possible to attain the goal towards which it moves, by gentler and kindlier ways. I expressed these views as forcibly as I then could in the above-mentioned book, with especial reference to improving the racial qualities of mankind, where the truest piety seems to me to reside in taking action, and not in submissive acquiescence to the routine of Nature. It was thought impious at one time to attach lightning conductors to churches, as showing a want of trust in the tutelary care of the Deity to whom they were dedicated; now I think most persons would be inclined to apply some contemptuous epithet to such as obstinately refused, on those grounds, to erect them.

The direct pursuit of studies in Eugenics, as to what could practically be done, and the amount of change in racial qualities that could reasonably be anticipated, did not at first attract investigators. The idea of effecting an improvement in that direction was too much in advance of the march of popular imagination, so I had to wait. In the meantime I occupied myself with collateral problems, more especially with that of dealing measurably with faculties that are variously distributed in a large population. The results were published in my 'Natural Inheritance' in 1889, and I shall have occasion to utilize some of them later on, in this very lecture. The publication of that book proved to be more timely than the former. The methods were greatly elaborated by Professor Karl Pearson, and applied by him to Biometry. Professor Weldon, of this University, whose untimely death is widely deplored, aided powerfully. A new science was thus created primarily on behalf of Biometry, but equally applicable to Eugenics, because their provinces overlap.

The publication of *Biometrika*, in which I took little more than a nominal part, appeared in 1901.

Being myself appointed Huxley Lecturer before the Anthropological Institute in 1901 I took for my title ' The possible improvement of the Human Breed under the existing conditions of Law and Sentiment ' *(Nature,* November 1, 1901, *Report of the Smithsonian Institute, Washington,* for the same year, and reprinted in this volume.)

The next and a very important step towards Eugenics was made by Professor Karl Pearson in his Huxley Lecture of 1903 entitled 'The Laws of Inheritance in Man' (*Biometrika*, vol. iii).   It contains a most valuable compendium of work achieved and of objects in view ; also the following passage (p. 159), which is preceded by forcible reasons for his conclusions :

We are ceasing as a nation to breed intelligence as we did fifty to a hundred years ago.   The mentally better stock in the nation is not reproducing itself at the same rate as it did of old; the less able and the less energetic are more fertile than the better stocks.   No scheme of wider or more thorough education will bring up, in the scale of intelligence, hereditary weakness to the level of hereditary strength.   The only remedy, if one be possible at all, is to alter the relative fertility of the good and the bad stocks in the community.

Again in 1904, having been asked by the newly-formed Sociological Society to contribute a memoir, I did so on 'Eugenics, its definition, aim and scope.' This was followed up in 1905 by three memoirs, 'Restrictions in Marriage,' 'Studies in National Eugenics,' and 'Eugenics as a factor in Religion,' which were published in the Memoirs of that Society with comments thereon by more than twenty different authorities *(Sociological Papers*, published for the Sociological Society (Macmillan), vols. i and ii.   These are re-published here).   The subject of Eugenics being thus formally launched, and the time appearing ripe, I offered a small endowment to the University of London, to found a Research Fellowship on its behalf. The offer was cordially accepted, so Eugenics gained the recognition of its importance by the University of London and a home for its study in University College. Mr. Edgar Schuster, of this University, became Research Fellow in 1905, and I am much indebted to his care in nurturing the young undertaking and for the memoirs he has contributed, part of which must still remain for a short time unpublished.

When the date for Mr. Schuster's retirement approached it was advisable to utilize the experience so far gained in reorganizing the Office. Professor Pearson and myself, in consultation with the authorities of the University of London, elaborated a scheme at the beginning of this year, which is a decided advance, and shows every sign of vitality and endurance. Mr. David Heron, a Mathematical Scholar of St. Andrew's, is now a Research Fellow; Miss Ethel Elderton, who has done excellent and expert work from the beginning, is deservedly raised to the position of Research Scholar; and the partial services of a trained Computer have been secured. An event of the highest importance to the future of the Office is that Professor Karl Pearson has undertaken, at my urgent request, that general supervision of its work which advancing age and infirmities preclude me from giving. He will, I trust, treat it much as an *annexe* to his adjacent biometric laboratory, for many studies in Eugenics might, with equal propriety, be carried on in either of them, and the same methods of precise analysis which are due to the mathematical skill and untiring energy of Professor Pearson are used in both. The Office now bears the name of the Eugenics Laboratory, and its temporary home is in 88 Gower Street. (It is now, 1909, housed in the University buildings.) The phrase ' National Eugenics ' is defined as ' the study of agencies under social control that may improve or impair the racial qualities of future generations, either physically or mentally.'

The Laboratory has already begun to publish memoirs on its own account, and I now rest satisfied in the belief that, with a fair share of good luck, this young Institution will prosper and grow into an important centre of research.

Application of Theories of Probability to Eugenics.

Eugenics seeks for quantitative results. It is not contented with such vague words as ' much ' or ' little,' but endeavours to determine ' how much ' or ' how little ' in precise and trustworthy figures. A simple example will show the importance of this. Let us suppose a class of persons, called A, who are afflicted with some form and some specified degree of degeneracy, as inferred from personal observations, and from family history, and let class B consist of the offspring of A. We already know only too well that when the grade of A is very low, that of the average B will be below par and mischievous to the community, but how mischievous will it probably be? This question is of a familiar kind, easily to be answered when a sufficiency of facts have been collected. But a second question arises, What will be the trustworthiness of the forecast derived from averages when it is applied to individuals? This is a kind of question that is not familiar, and rarely taken into account, although it too could be answered easily as follows. The average mischief done by each B individual to the community may for brevity be called M: the mischiefs done by the several individuals differ more or less from M by amounts whose average may be called D. In other words D is the average amount of the individual deviations from M. D thus becomes the measure of untrustworthiness. The smaller D is, the more precise the forecast, and the stronger the justification for taking such drastic measures against the propagation of class B as would be consonant to the feelings if the forecast were known to be infallible. On the other hand, a large D signifies a corresponding degree of uncertainty, and a risk that might be faced without reproach through a sentiment akin to that expressed in the maxim ' It is better that many guilty should escape than that one innocent person should suffer.' But that is not the sentiment by which natural selection is guided, and it is dangerous to yield far to it.

There can be no doubt that a thorough investigation of the kind described, even if confined to a single grade and to a single form of degeneracy, would be a serious undertaking. Masses of trustworthy material must be collected, usually with great difficulty, and be afterwards treated with skill and labour by methods that few at present are competent to employ. An extended investigation into the good or evil done to the State by the offspring of many different classes of persons, some of civic value, others the reverse, implies a huge volume of work sufficient to occupy Eugenics laboratories for an indefinite time.

OBJECT LESSONS IN THE METHODS OF BIOMETRY.

I propose now to speak of those fundamental principles of the laws of Probability that are chiefly concerned in the newer methods of Biometry, and consequently of Eugenics. Most persons of ordinary education seem to know nothing about them, not even understanding their technical terms, much less appreciating the cogency of their results. This popular ignorance so obstructs the path of Eugenics that I venture to tax your attention by proposing a method of partly dispelling it. Let me first say that no one can be more conscious than myself of the large amount of study that is required to qualify a man to deal adequately with the mathematical methods of Biometry, or that any man can hope for much success in that direction unless he is possessed of appropriate faculties and a strong brain. On the other hand, I hold an opinion likely at first sight to scandalize biometricians and which I must justify, that the fundamental ideas on which abstruse problems of Probability are based admit of being so presented to any intelligent person as to be grasped by him, even though he be quite ignorant of mathematics. The conditions of doing so are that the lessons shall be as far as possible 'Object lessons,' in which real objects shall be handled as in the Kindergarten system, and simple operations performed and not only talked about. I am anxious to make myself so far understood, that some teachers of science may be induced to elaborate the course that I present now only in outline. It seems to me suitably divisible into a course of five lessons of one hour each, which would be sufficient to introduce the learner into a new world of ideas, extraordinarily wide in their application. A proper notion of what is meant by Correlation requires some knowledge of the principal features of Variation, and will be the goal towards which the lessons lead.

To most persons Variability implies some-

thing indefinite and capricious.   They require
to be taught that it, like Proteus in the old fable,
can be seized, securely bound, and utilized ;
that it can be defined and measured.   It was
disregarded by the old methods of statistics,
that concerned themselves solely with Averages.   The average amount of various
measurable faculties or events in a multitude
of persons was determined by simple methods,
the individual variations being left out of
account as too difficult to deal with. A population was treated by the old methods as a
structureless atom, but the newer methods
treat it as a compound unit. It will be a considerable intellectual gain to an otherwise
educated person, to fully understand the way
in which this can be done, and this and such
like matters the proposed course of lessons
is intended to make clear. It cannot be
expected that in the few available minutes
more than an outline can be given here of
what is intended to be conveyed in perhaps
thirty-fold as much time with the aid of pro-
fuse illustrations by objects and diagrams. At
the risk of being wearisome, it is, however,
necessary to offer the following syllabus of
what is proposed, for an outline of what
teachers might fill in.

The object of the first lesson would be to
explain and illustrate Variability of Size,
Weight, Number, &c., by exhibiting samples
of specimens that have been marshalled at
random (Fig. 1), or arrayed in order of their

magnitude (Fig. 2). Thus when variations of length were considered, objects of suitable size, such as chestnuts, acorns, hazel-nuts, stones of wall fruit, might be arrayed as beads on a string. It will be shown that an 'Array' of Variates of any kind falls into a continuous series. That each variate differs little from its neighbours about the middles of the Arrays, but that such differences increase rapidly towards their extremities. Abundant illustration would be required, and much handling of specimens.

Arrays of Variates of the same class strung together, differing considerably in the number of the objects they each contain, would be laid side by side and their middlemost variates or 'Medians' (Fig. 3) would be compared. It would be shown that as a rule the Medians become very similar to one another when the numbers in the Arrays are large. It must then be dogmatically explained that double accuracy usually accompanies a four-fold number, treble accuracy a nine-fold number, and so on.

(This concludes the first lesson, during which the words and significations of Variability, Variate, Array, and Median will have been learnt.)

The second lesson is intended to give more precision to the idea of an Array. The variates in any one of these strung loosely on a cord, should be disposed at equal distances apart in front of an equal number of compartments, like horses in the front of a row of stalls (Fig. 4), and their tops joined.   There will be one more side to the row of stalls than there are horses, otherwise a side of one of the extreme stalls would be wanting.   Thus there are two ways of indicating the position of a particular variate, either by its *serial number* as ' first,' ' second,' ' third,' or so on, or by *degrees* like those of a thermometer. In the latter case the sides of the stalls serve as degrees, counting the first of them as 0°, making one more graduation than the number of objects, as it should be.   The difference between these two methods has to be made clear, and that while the serial position of the Median object is always the same in any two Arrays whatever be the number of variates, the serial position of their subdivisions cannot be the same, the ignored half interval at either end varying in width according to the number of variates, and becoming considerable when that number is small.

Lines of proportionate length will then be drawn on a blackboard, and the limits of the Array will be also drawn, at a half interval from either of its ends.   The base is then to be divided centesimally.

Next join the tops of the lines with a smooth curve, and wipe out everything except the curve, the Limit at either side, and the Centesimally divided Base (Fig. 5).   This figure forms a Scheme of Distribution of Variates.   Explain clearly that its shape is independent of the number of Variates, so long as they are sufficiently numerous to secure statistical constancy.

Show numerous schemes of variates of different kinds, and remark on the prevalent family likeness between the bounding curves. (Words and meanings learnt—Schemes of Distribution, Centesimal graduation of base.)

The third lesson passes from Variates, measured upwards from the base, to Deviates measured upwards or downwards from the Median, and treated as positive or negative values accordingly (Fig. 6).

Draw a Scheme of Variates on the blackboard, and show that it consists of two parts ; the median which represents a constant, and the curve which represents the variations from it. Draw a horizontal line from limit to limit, through the top of the Median to serve as Axis to the Curve. Divide the Axis centesimally, and wipe out everything except Curve, Axis, and Limits. This forms a Scheme of Distribution of Deviates. Draw ordinates from the axis to the curve at the 25th and 75th divisions. These are the ' Quartile ' deviates.

At this stage the Genesis of the theoretical Normal curve might be briefly explained and the generality of its application ; also some of its beautiful properties of reproduction. Many of the diagrams already shown would be again employed to show the prevalence of approximately normal distributions. Exceptions of strongly marked Skew curves would be exhibited and their genesis briefly described.

It will then be explained that while the ordinate at *any* specified centesimal division in two normal curves of deviation measures their relative variability, the Quartile is commonly employed as the unit of variability under the almost grotesque name of 'Probable Error,' which is intended to signify that the length of anyDeviate in the system is as likely as not to exceed or to fall short of it. This, by construction, is the case of either Quartile.

(New words and meanings—Scheme of Distribution of Deviates, Axis, Normal, Skew, Quartile, and Probable Error.)

In the fourth lesson it has to be explained that the Curve of Normal Distribution is not a direct result of calculation, neither does the formula that expresses it lend itself so freely to further calculation, as the curve of Frequency. Their shapes differ ; the first is an Ogive, the second (Fig. 7) is Bell-shaped. In the curve of Frequency the Deviations are reckoned from the Mean of all the Variates, and not from the Median. Mean and Median are the same in Normal Curves, but may differ much in others. Either of these normal curves can be transformed into the other, as is best exemplified by using a Polygon (Fig. 8) instead of the Curve, consisting of a series of rectangles differing in height by the same amounts, but having widths respectively representative of the frequencies of 1, 3, 3, 1.   (This is one of those known as a Binomial series, whose genesis might be briefly explained.)   If these rectangles are arrayed in order of their widths, side by side, they become the equivalents of the ogival curve of Distribution.   Now if each of these latter rectangles be slid parallel to itself up to either limit, their bases will overlap and they become equivalent to the bell-shaped curve of Frequency with its base vertical.

The curve of Frequency contains no easily perceived unit of variability like the Quartile of the Curve of Distribution.   It is therefore not suited for and was not used as a first illustration, but the formula that expresses it is by far the more suitable of the two for calculation.   Its unit of variability is what is called the 'Standard Deviation,' whose genesis will admit of illustration.   How the calculations are made for finding its value is far beyond the reach of the present lessons.   The calculated ordinates of the normal curve must be accepted by the learner much as the time of day by his watch, though he be ignorant of the principles of its construction.   Much further beyond his reach are the formulae used to express quasi-normal and skew curves.   They require a previous knowledge of rather advanced mathematics.

(New words and ideas—Curve of Frequency, Standard Deviation, Mean, Binomial Series).

The fifth and last lesson deals with the measurement of Correlation, that is, with the closeness of the relation between any two systems whose variations are due partly to causes common to both, and partly to causes special to each. It applies to nearly every social relation, as to environment and health, social position and fertility, the kinship of parent to child, of uncle to nephew, &c. It may be mechanically illustrated by the movements of two pulleys with weights attached, suspended from a cord held by one of the hands of three different persons, 1, 2, and 3. No. 2 holds the middle of the cord, one half of which then passes round one of the pulleys up to the hand of No. 1 ; the other half similarly round the other pulley up to the hand of No. 3. The hands of Nos. 1, 2, and 3 move up and down quite independently, but as the movements of both weights are simultaneously controlled in part by No. 2, they become ' correlated.'

The formation of a table of correlations on paper ruled in squares, is easily explained on the blackboard (Fig. 9). The pairs of correlated values A and B have to be expressed in units of their respective variabilities. They are then sorted into the squares of the paper, —vertically according to the magnitudes of A, horizontally according to those of B—,and the Mean of each partial array of B values, corresponding to each grade of A, has to be determined. It is found theoretically that where variability is normal, the Means of B lie practically in a straight line on the face of the Table, and observation shows they do so in most other cases. It follows that the average deviation of a B value bears a constant ratio to the deviation of the corresponding A value. This ratio is called the 'Index of Correlation,' and is expressed by a single figure. For example: if the thigh-bone of many persons deviate 'very much' from the usual length of the thigh-bones of their race, the average of the lengths of the corresponding arm-bones will differ 'much,' but not 'very much,' from the usual length of arm-bones, and the ratio between this 'very much' and 'much' is constant and in the same direction, whatever be the numerical value attached to the word 'very much.' Lastly, the trustworthiness of the Index of Correlation, when applied to individual cases, is readily calculable. When the closeness of correlation is absolute, it is expressed by the number 1·0; and by 0·0, when the correlation is nil.

(New words and ideas—Correlation and Index of Correlation.)

This concludes what I have to say on these suggested Object lessons. It will have been tedious to follow in its necessarily much compressed form,—but will serve, I trust, to convey its main purpose of showing that a very brief course of lessons, copiously illustrated by diagrams and objects to handle, would give an acceptable introduction to the newer methods employed in Biometry and in Eugenics. Further, that when read leisurely by experts in its printed form, it would give them sufficient guidance for elaborating details.

INFLUENCE OF COLLECTIVE TRUTHS UPON INDIVIDUAL CONDUCT.

We have thus far been concerned with Probability, determined by methods that take cognizance of Variations, and yield exact results, thereby affording a solid foundation for action. But the stage on which human action takes place is a superstructure into which emotion enters, we are guided on it less by Certainties and by Probabilities than by Assurance to a greater or lesser degree. The word Assurance is derived from *sure*, which itself is an abbreviation of *secure*, that is of *se- cura*, or without misgiving. It is a contented attitude of mind largely dependent on custom, prejudice, or other unreasonable influences which reformers have to overcome, and some of which they are apt to utilize on their own behalf. Human nature is such that we rarely find our way by the pure light of reason, but while peering through spectacles furnished with coloured and distorting glasses.

Locke seems to confound certainty with assurance in his forcible description of the way in which men are guided in their daily affairs (*Human Understanding*, iv. 14, par. 1):

Man would be at a great loss if he had nothing to direct him but what has the certainty of true knowledge. For that being very short and scanty, he would be often utterly in the dark, and in most of the actions of his life, perfectly at a stand, had he nothing to guide him in the absence of clear and certain knowledge. He that will not eat till he has demonstration that it will nourish him, he that will not stir till he infallibly knows the business he goes about will succeed, will have little else to do than to sit still and perish.

A society may be considered as a highly complex organism, with a consciousness of its own, caring only for itself, establishing regulations and customs for its collective advantage, and creating a code of opinions to subserve that end. It is hard to over-rate its power over the individual in regard to any obvious particular on which it emphatically insists. I trust in some future time that one of those particulars will be the practice of Eugenics. Otherwise the influence of collective truths on individual conduct is deplorably weak, as expressed by the lines :—

> For others' follies teach us not,
> Nor much their wisdom teaches,
> But chief of solid worth is what
> Our own experience preaches.

Professor Westermarck, among many other remarks in which I fully concur, has aptly stated *(Sociological Papers*, published for the Sociological Society. Macmillan, 1906, **vol.** ii., p. 24), with reference to one obstacle which prevents individuals from perceiving the importance of Eugenics, ' the prevalent opinion that almost anybody is good enough to marry is chiefly due to the fact that in this case, cause and effect, marriage and the feebleness of the offspring, are so distant from each other that the *near-sighted eye* does not distinctly perceive the connexion between them.'   (The Italics are mine.)

The enlightenment of individuals is a necessary preamble to practical Eugenics, but social opinion is the tyrant by whose praise or blame the principles of Eugenics may be expected hereafter to influence individual conduct.   Public opinion may, however, be easily directed into different channels by opportune pressure.   A common conviction that change in the established order of some particular codes of conduct would be impossible, because of the shock that the idea of doing so gives to our present ideas, bears some resemblance to the conviction of lovers that their present sentiments will endure for ever. Conviction, which is that very Assurance of which mention has just been made, is proved by reiterated experience to be a ‚highly fallacious guide.   Love is notoriously fickle in despite of the fervent and genuine protestations of lovers, and so is public opinion. I gave a list of extraordinary variations of the latter in respect to restrictions it enforced on the freedom of marriage, at various times and places *(Sociological Papers*, quoted above). Much could be added to that list, but I will not now discuss the effects of public opinion on such a serious question. I will take a much smaller instance which occurred before the time to which the recollections of most persons can now reach, but which I myself recall vividly. It is the simple matter of hair on the face of male adults. When I was young, it was an unpardonable offence for any English person other than a cavalry officer, or perhaps someone of high social rank, to wear a moustache. Foreigners did so and were tolerated, otherwise the assumption of a moustache was in popular opinion worse than wicked, for it was atrociously bad style. Then came the Crimean War and the winter of Balaclava, during which it was cruel to compel the infantry to shave themselves every morning. So their beards began to grow, and this broke a long established custom. On the return of the army to England the fashion of beards spread among the laity, but stopped short of the clergy. These, however, soon began to show dissatisfaction; they said the beard was a sign of manliness that ought not to be suppressed, and so forth, and at length the moment arrived. A distinguished clergyman, happily still living, 'bearded' his Bishop on a critical occasion, The Bishop yielded without protest, and forthwith hair began to sprout in a thousand pulpits where it had never appeared before within the memory of man.

It would be no small shock to public

sentiment if our athletes in running public races were to strip themselves stark naked, yet that custom was rather suddenly introduced into Greece. Plato says (Republic V, par. 452, Jowett's translation) :

Not long ago the Greeks were of the opinion, which is still generally received among the barbarians, that the sight of a naked man was ridiculous and improper, and when first the Cretans and the Lacedaemonians introduced naked exercises, the wits of that day might have ridiculed them. . . .

Thucydides (I. 6) also refers to the same change as occurring ' quite lately '.

Public opinion is commonly far in advance of private morality, because society as a whole keenly appreciates acts that tend to its advantage, and condemns those that do not. It applauds acts of heroism that perhaps not one of the applauders would be disposed to emulate. It is instructive to observe cases in which the benevolence of public opinion has out-stripped that of the Law—which, for example, takes no notice of such acts as are enshrined in the parable of the good Samaritan. A man on his journey was robbed, wounded and left by the wayside. A priest and a Levite successively pass by and take no heed of him. A Samaritan follows, takes pity, binds his wounds, and bears him to a place of safety. Public opinion keenly condemns the priest and the Levite, and praises the Samaritan, but our criminal law is indifferent to such acts. It is most severe on misadventure due to the neglect of a definite duty, but careless about those due to the absence of common philanthropy. Its callousness in this respect is painfully shown in the following quotations (Kenny, *Outlines of Criminal Law*, 1902, p. 121, per Hawkins in Reg. v. Paine, *Times*, February 25, 1880):

If I saw a man who was not under my charge, taking up a tumbler of poison, I should not be guilty of any crime by not stopping him. I am under no legal obligation to protect a stranger.

That is probably what the priest and the Levite of the parable said to themselves.

A still more emphatic example is in the *Digest of Criminal Law*, by Justice Sir James Stephen, 1887, p. 154. Reg. v. Smith, 2 C and P., 449:

A sees *B* drowning and is able to help him by holding out his hand. *A* abstains from doing so in order that *B* may be drowned, and *B* is drowned. *A* has committed no offence.

It appears, from a footnote, that this case has been discussed in a striking manner by Lord Macaulay in his notes on the Indian Penal Code, which I have not yet been able to consult.

Enough has been written elsewhere by myself and others to show that whenever public opinion is strongly roused it will lead to action, however contradictory it may be to previous custom and sentiment. Considering that public opinion is guided by the sense of what best serves the interests of society as a whole, it is reasonable to expect that it will be strongly exerted in favour of Eugenics when a sufficiency of evidence shall have been collected to make the truths on which it rests plain to all.   That moment has not yet arrived.   Enough is already known to those who have studied the question to leave no doubt in their minds about the general results, but not enough is quantitatively known to justify legislation or other action except in extreme cases.   Continued studies will be required for some time to come, and the pace must not be hurried.   When the desired fulness of information shall have been acquired then, and not till then, will be the fit moment to proclaim a ' Jehad,' or Holy War against customs and prejudices that impair the physical and moral qualities of our race.

LOCAL ASSOCIATIONS FOR PROMOTING EUGENICS*

I propose to take the present opportunity of submitting some views of my own relating to that large province of eugenics which is concerned with favouring the families of those who are exceptionally fit for citizenship. Consequently, little or nothing will be said relating to what has been well termed by Dr. Saleeby "negative" eugenics, namely, the hindrance of the marriages and the production of offspring by the exceptionally unfit. The latter is unquestionably the more pressing subject of the two, but it will soon be forced on the attention of the legislature by the recent report of the Royal Commissio.1 on the Feeble-minded. We may be content to await for awhile the discussions to which it will give rise, and which I am sure the members of this society will follow with keen interest, and with readiness to intervene when what may be advanced seems likely to result in actions of an anti-eugenic character.

The remarks I am about to make were suggested by hearing of a desire to further eugenics by means of local associations more or less affiliated to our own, combined with

---

* Address to a meeting of the Eugenics Education Society at the Grafton Galleries, on October 14th, 1908.

much doubt as to the most appropriate methods of establishing and conducting them. It is upon this very important branch of our wide subject that I propose to offer a few remarks.

It is difficult, while explaining what I have in view, to steer a course that shall keep clear of the mud flats of platitude on the one hand, and not come to grief against the rocks of over-precision on the other. There is no clear issue out of mere platitudes, while there is great danger in entering into details. A good scheme may be entirely compromised merely on account of public opinion not being ripe to receive it in the proposed form, or through a discovered flaw in some non-essential part of it. Experience shows that the safest course in a new undertaking is to proceed warily and tentatively towards the desired end, rather than freely and rashly along a predetermined route, however carefully it may have been elaborated on paper.

Again, whatever scheme of action is proposed for adoption must be neither Utopian nor extravagant, but accordant throughout with British sentiment and practice.

The successful establishment of any general system of constructive eugenics will, in my view (which I put forward with diffidence), depend largely upon the efforts of local associations acting in close harmony with a central society, like our own. A prominent part of its business will then consist in affording opportunities for the interchange of ideas and for the registration and comparison of results. Such a central society would tend to bring about a general uniformity of administration the value of which is so obvious that I do not stop to insist on it.

Assuming, as I do, that the powers at the command of the local associations will be almost purely social, let us consider how those associations might be formed and conducted so as to become exceedingly influential.

It is necessary to be somewhat precise at the outset, so I will begin with the by no means improbable supposition that in a given district a few individuals, some of them of local importance, are keenly desirous of starting a local association or society, and are prepared to take trouble to that end. How should they set to work?

Their initial step would seem to be to form themselves into a provisional executive committee, and to nominate a president, council, and other officers of the new society. This done, the society in question, though it would have no legal corporate existence, may be taken as formed.

The committee would next provide, with the aid of the central society, for a few sane and sensible lectures to be given on eugenics, including the A B C of heredity, at some convenient spot, and they would exert themselves to arouse a wide interest in the subjects by making it known in the district. They would seek the co-operation of the local medical men, clergy, and lawyers, of the sanitary authorities, and of all officials whose administrative duties bring them into contact with various classes of society, and they would endeavour to collect round this nucleus that portion of the local community which was likely to be brought into sympathy with the eugenic cause. Every political organisation, every philanthropic agency, proceeds on some such lines as I have just sketched out.

The committee might next issue, on the part of the president and council of the new society, a series of invitations to guests at their social gatherings, where differences of rank should be studiously ignored. The judicious management of these gatherings would, of course, require considerable tact, but there are abundant precedents for them, among which I need only mention the meetings of the Primrose League at one end of the scale, and those held in Toynbee Hall at the other end. Given a not inclement day, an hour suitable to the occasion, a park or large garden to meet in, these informal yet select reunions might be made exceedingly pleasant, and very helpful to the eugenic cause.

The inquiries made by the committee when they were considering the names of strangers to whom invitations ought to be sent, would put them in possession of a large fund of information concerning the qualities of many notable individuals in their district, and their family histories. These family histories should be utilised for eugenic studies, and it should be the duty of the local council to cause them to be tabulated in an orderly way, and to communicate the more significant of them to the central society.

The chief of the notable qualities, to which I refer in the preceding paragraph, is the possession of what I will briefly call by the general term of " Worth." By this I mean the civic worthiness, or the value to the State, of a person, as it would probably be assessed by experts, or, say, by such of his fellow-workers as have earned the respect of the community in the midst of which they live. Thus the worth of soldiers would be such as it would be rated by respected soldiers, students by students, business men by business men, artists by artists, and so on. The State is a vastly complex organism, and the hope of obtaining a proportional representation of its best parts should be an avowed object of issuing invitations to these gatherings.

Speaking only for myself, if I had to classify persons according to worth, I should consider each of them under the three heads of physique, ability, and character, subject to the provision that inferiority in any one of the three should outweigh superiority in the other two. I rank physique first, because it is not only very valuable in itself and allied to many other good qualities, but has the additional merit of being easily rated. Ability I should place second on similar grounds, and character third, though in real importance it stands first of all. It is very difficult to rate character justly; the tenure of a position of trust is only a partial test of it, though a good one so far as it goes. Again, I wish to say emphatically that in what I have thrown out I have no desire to impose my own judgment on others, especially as I feel persuaded that almost any intelligent committee would so distribute their invitations to strangers as to include most, though perhaps not all, of the notable persons in the district.

By the continued action of local associations as described thus far, a very large amount of good work in eugenics would be incidentally done. Family histories would become familiar topics, the existence of good stocks would be discovered, and many persons of "worth" would be appreciated and made acquainted with each other who were formerly known only to a very restricted circle. It is probable that these persons, in their struggle to obtain appointments, would often receive valuable help from local sympathisers with eugenic principles. If local societies did no more than this for many years to come, they would have fully justified their existence by their valuable services.

A danger to which these societies will be liable arises from the inadequate knowledge joined to great zeal of some of the most active among their probable members. It may be said, without mincing words, with regard to much that has already been published, that the subject of eugenics is particularly attractive to " cranks." The councils of local societies will therefore be obliged to exercise great caution before accepting the memoirs offered to them, and much discretion in keeping discussions within the bounds of sobriety and common sense. The basis of eugenics is already firmly established, namely, that the offspring of "worthy" parents are, *on the whole*, more highly gifted by nature with faculties that conduce to " worthiness " than the offspring of less " worthy " parents. On the other hand, forecasts in respect to particular cases may be quite wrong. They have to be based on imperfect data. It cannot be too emphatically repeated that a great deal of careful statistical work has yet to be accomplished before the science of eugenics can make large advances.

I hesitate to speculate farther. A tree will have been planted ; let it grow. Perhaps those who may thereafter feel themselves or be considered by others to be the possessors of notable eugenic qualities—let us for brevity call them " Eugenes "—will form their own clubs and look after their own interests. It is impossible to foresee what the state of public opinion will then be. Many elements of strength are needed, many dangers have to be evaded or overcome, before associations of Eugenes could be formed that would be stable in themselves, useful as institutions, and approved of by the outside world.

The suggestion I made in the earlier part of this paper that the executive committee of local associations should co-operate, wherever practicable, with local administrative authorities, proceeded on the assumption that the inhabitants of the districts selected as the eugenic "field" had a public spirit of their own and a sense of common interest. This sense would be greatly strengthened by the enlargement of mutual acquaintanceship and the spread of the eugenic idea consequent on the tactful action of the committee. It ought not to be difficult to arouse in the inhabitants a just pride in their own civic worthiness, analogous to the pride which a soldier feels in the good reputation of his regiment or a lad in that of his school. By this means a strong local eugenic opinion might easily be formed. It would be silently assisted by local object lessons, in which the benefits derived through following eugenic rules and the bad effects of disregarding them were plainly to be discerned.

The power of social opinion is apt to be underrated rather then overrated. Like the atmosphere which we breathe and in which we move, social opinion operates powerfully without our being conscious of its weight. Everyone knows that governments, manners, and beliefs which were thought to be right, decorous, and true at one period have been judged wrong, indecorous, and false at another ; and that views which we have heard expressed by those in authority over us in our childhood and early manhood tend to become axiomatic and unchangeable in mature life.

In circumscribed communities especially, social approval and disapproval exert a potent force. Its presence is only too easily read by those who are the object of either, in the countenances, bearing, and manner of persons whom they daily meet and converse with. Is it, then, I ask, too much to expect that when a public opinion in favour of eugenics has once taken sure hold of such communities and has been accepted by them as a quasi-religion, the result will be manifested in sundry and very effective modes of action which are as yet untried, and many of them even unforeseen ?

Speaking for myself only, I look forward to local eugenic action in numerous directions, of which I will now specify one. It is the accumulation of considerable funds to start young couples of " worthy " qualities in their married life, and to assist them and their families at critical times. The gifts to those who are the reverse of "worthy" are enormous in amount; it is stated that the charitable donations or bequests in the year 1907 amounted to 4,868,050*l.* I am not prepared to say how much of this was judiciously spent, or in what ways, but merely quote the figures to justify the inference that many of the thousands of persons who are willing to give freely at the prompting of a sentiment based upon compassion might be persuaded to give largely also in response to the more virile desire of promoting the natural gifts and the national efficiency of future generations.